Ecology for Gardeners

Ecology for Gardeners

Steven B. Carroll and Steven D. Salt

Drawings by Errol D. Hooper Jr.

TIMBER PRESS
Portland ▪ Cambridge

Published in 2004 by
Timber Press, Inc.
The Haseltine Building
133 S.W. Second Avenue, Suite 450
Portland, Oregon 97204, U.S.A.

Timber Press
2 Station Road
Swavesey
Cambridge CB4 5QJ, U.K.

Printed in China

Library of Congress Cataloging-in-Publication Data

Carroll, Steven B., 1952–
    Ecology for gardeners / Steven B. Carroll and Steven D. Salt.
        p. cm.
Includes bibliographical references (p.   ).
    ISBN 0-88192-611-6
    1. Garden ecology.   I. Salt, Steven D., 1952– II. Title.
QH541.5.G37 C27 2004
577.5'54--dc21

A catalog record for this book is also available from the British Library.

To The Nature Conservancy and Seed Savers Exchange,
for preserving the world's natural and gardening
resources one habitat and one seed at a time.

And to our families, without whom we would have
neither this book nor our gardens.

Contents

Plates follow page 96

Acknowledgments

In the course of researching and writing this book, we sought ideas and advice from others at every turn, and we leaned heavily on friends, colleagues, and anyone else who strayed too close. So, it is with great pleasure that we thank the many people who helped make the book a reality.

For providing photographs or electron micrographs, we thank Al Cornell, Pete Goldman, Brock Neil, William Sanders, Kent Whealy (Seed Savers Exchange), Kristin Hay and Merlin Tuttle (Bat Conservation International), Mike Klawitter and Liz Petrus (Gempler's, Inc.), and Anita Daniels, Bill Tarpening, and Vivian Thomas (U.S. Department of Agriculture). We are especially grateful to Anne Bergey, Michael Kelrick, Jeff Osborn, and George Shinn, who not only provided many images used in this book but gave us free access to the thousands of slides in their personal collections. We particularly enjoyed working with Errol Hooper, whose superb illustrations add so much to this book. We greatly appreciate the photographic advice of Mark Gray, at Northeast Camera and Photo Labs, and Tim Barcus, Truman State University campus photographer. For assistance with greenhouse plantings, we thank Amy Miller.

We thank Alisha Gammon, Lisa Hooper, and Christine Ritchie for providing helpful comments on sections of the book; Lisa also contributed her botanical expertise during the illustration process. Paula Presley provided much-needed guidance in developing the indexes. Cheryl Gracey read the entire manuscript and massaged the prose of two academics into something considerably more readable—and for this she was paid only in chocolate. Dale Johnson and Lisa DiDonato, our editors at Timber Press, have been a great help and paragons of patience as this project went on longer than we (and perhaps they) ever imagined. Lisa's knowledge of the natural world has added immeasurably to this book.

Large projects have a way of taking on a life of their own. As we spent ever more time at our keyboards, our families helped when they could, did more than their share of weeding in the garden, and waited patiently for the words, "The End." For these things, we thank them.

Prologue

This book is about gardening. It is also about ecology, the branch of science that deals with how organisms interact with each other and with their environment. Nature brings together sunlight, air, water, minerals, and a dazzling array of organisms to form forests, marshes, grasslands, and other natural ecosystems. Gardeners, with the help of tools, take advantage of and manipulate the same raw materials to create gardens. Thus, gardens are simply human-managed ecosystems.

Because everything in an ecosystem is intricately interconnected, deciding how to organize this book was like asking where the beginning of a circle is, or which piece of a jigsaw puzzle comes next. There is no single correct way to organize such a book. We hope, however, that our scheme makes sense and that it will help gardeners understand some of the principles of ecology and how to apply them to gardening.

First, we look at the most prominent actors in garden ecosystems—plants—especially considering ecologically relevant attributes such as structure, development, genetics, and reproduction as well as taxonomic conventions. In chapter 2 we introduce other organisms that visit or inhabit gardens. We describe in greatest detail those creatures that seem to have the most influential roles, such as insects, nematodes, and microorganisms, but also consider other garden denizens ranging from elephants to prions, some of which may be unfamiliar to most gardeners. Next we explore sunlight, air, waters, and the soil—the physical environment within which garden organisms live and function. In chapter 4 we consider the ways plants adapt to and function in their physical environment and how they come together in communities. We then examine the intricate interactions among garden-dwelling organisms, ranging from mutually harmful to mutually beneficial. In chapter 6 we introduce humans into the garden ecosystem and consider how gardeners can use ecological principles in wisely managing their garden ecosystems. We have tried to minimize the use of scientific jargon as far as possible. However, you may still encounter unfamiliar words. Peruse the book's extensive glossary, expand your vocabulary, become a more knowledgeable gardener, and maybe win a few more Scrabble games.

Prologue

As far as possible, we have included examples that are directly applicable to gardens and gardening. However, most ecologists have carried out their scientific studies in natural ecosystems—those created by Nature with no help from humans. Thus, we have also drawn many examples from these natural systems to illustrate ecological principles. After all, the same ecological principles at work in, say, Canadian spruce-fir forests are found in gardens; these principles simply may not have been as well investigated in gardens. When left alone, Nature seems to do a pretty good job of managing ecosystems. It would be wise, then, for gardeners to study natural processes to gain some tips for wisely managing garden ecosystems. And that is what this book is all about.

1

The Nature
of Plants

Plants comprise more than a quarter million species and are found in virtually every habitat on Earth. To survive in environments as different as the hot deserts of North Africa, the windswept tundra of the Arctic, and the coniferous rainforests of the Pacific Northwest, plants employ a remarkable diversity of structural, anatomical, and physiological adaptations (plate 1). Plants growing side by side in a garden are not subject to such dramatic variation in environmental conditions, but sunlight, air temperature at ground level, soil chemistry, and countless other factors can vary to a surprising extent over short distances. As a result, plants having different requirements for optimal growth and reproduction do best in different parts of the same garden.

The Plant Body

Most plants grown as ornamentals or food crops are flowering plants (angiosperms). Plants in this group characteristically produce leaves, stems, and roots during vegetative growth and flowers, fruits, and seeds during the sexual portion of the life cycle. Other kinds of plants, such as conifers and mosses (bryophytes), produce some, but not others, of these structures. By understanding the structure and function of these plant tissues, gardeners can make better-informed decisions about such things as plant selection, cultivation, and pest management.

Leaves

In many plants leaves are the most conspicuous organs, at least during some parts of the year. Among broad-leaf plants, most leaves have a distinct, flattened blade and a stalked petiole by which the blade attaches to the stem. Blade characteristics affect traits such as the amount of light intercepted by a particular leaf and the amount of light reaching lower leaves, which in turn affects the efficiency with which a plant harnesses solar energy.

Leaves come in a dazzling variety of sizes, shapes, colors, textures, and arrangements along the stem. They range in size from less than $^4/_{100}$ inch (1 mm) wide in aquatic water meal (*Wolffia*) to more than 20 feet (6 m) long in the Seychelles palm (*Lodoicea maldivica*). Leaves may be thick and waxy to prevent water loss, as in *Agave* and *Magnolia*; feathery and delicate, as in dill (*Anethum graveolens*); or soft, as in velvet-leaf (*Abutilon theophrasti*); and they may be virtually any thickness and texture in between. Most leaves are a shade of green, but colors such as red, purple, and yellow are also common. Indeed, leaf color has been purposely manipulated by plant geneticists and growers to achieve aesthetic effects.

Leaf characteristics greatly influence the choice of garden and lawn plantings. Hostas (*Hosta*) are grown primarily for their diversity of solid, striped, and variegated leaves, whereas ferns, which often share shaded sites with hostas, are usually grown for their textured, dissected leaves. Ground covers such as pachysandra (*Pachysandra terminalis*) and common periwinkle (*Vinca minor*) are frequently favored in shaded, dry, and erosion-prone sites for their glossy, evergreen leaves.

Leaves are sometimes modified in striking ways, as in the sticky, insect-trapping leaves of sundew (*Drosera*) and the twining tendrils of sweet pea (*Lathyrus odoratus*). From an ecological perspective, tendrils and other means of attachment allow plants to grow toward light without producing energetically costly supportive tissues such as stems. Climbing plants such as grapes (*Vitis*) are encouraged in and around the garden; but they may also be reviled, for example, when field bindweed (*Convolvulus arvensis*) climbs up and over plantings.

Simple leaves are composed of a single flattened blade, for example, those of *Caladium* (plate 2) and wild ginger (*Asarum*), whereas in compound leaves blades are subdivided into multiple leaflets, such as those of sumac (*Rhus*) and elderberry (*Sambucus*). The margin of leaf blades may be smooth, wavy, notched, saw-toothed, and virtually any other shape imaginable. This aspect of leaves affects the shedding of water, the capture of light, and the ease with which herbivores recognize and feed on leaves. Leaf behavior also influences plant selection by growers, with the sudden folding of the leaves of sensitive plant (*Mimosa pudica*) in response to touch or a strong wind representing a particularly dramatic example.

Nodes are the points along the stem at which leaves and branches arise, and the arrangement of leaves at nodes varies substantially (plate 3). In lupines (*Lupinus*) and black-eyed Susan (*Rudbeckia hirta*), each node gives rise to only a single leaf. In other plants, two or more leaves may form at the same node, as in maple (*Acer*) and *Trillium*.

A leaf's petiole connects the blade to the stem and contains the conducting tissue that transports water and dissolved substances, including the sugars synthesized in photosynthesis (plate 4). In some plants, the petiole can bend to move the leaf into higher-light areas. Although the petiole is usually not a focus of garden and lawn plantings, a feature of the well-named quaking aspen (*Populus tremuloides*) is a fluttering of its leaves in even slight breezes; this quaking happens when the flattened petioles catch the wind. Variation in leaf petioles is useful in plant identification, and it often imparts a distinctive visual quality in the garden.

Conducting tissues that enter the leaf blade form veins, which are usually more prominent on the underside of a leaf. Many leaves have a large, central midvein and increasingly smaller branching veins that reach all parts of the leaf interior. A slight or sometimes substantial color difference between the leaf-blade surface and veins can create a pleasing contrast in the leaves of garden plants, as seen in Persian shield (*Strobilanthes dyerianus*).

Not all leaves have flat, broad blades, however. Grasses (family Poaceae or Gramineae), lilies (*Lilium*), and many other monocots have relatively narrow leaves whose major, conspicuous veins run parallel to each other (plate 5). Conifers (phylum Coniferophyta) such as yews (*Taxus*) and firs (*Abies*) generally have needle-shaped or flat, narrow leaves. Most conifer leaves remain on the plant for anywhere from one to twenty years, depending on the species and conditions, although the leaves of larch (*Larix*) and bald cypress (*Taxodium distichum*) turn yellow and drop off each autumn. Conifer leaves have many other internal and external features that differ from leaves of flowering plants; for example, they are usually needle-shaped or scalelike and have resin canals and sunken stomates.

One of the chief features of plants is the ability to carry out photosynthesis—a series of chemical reactions in which carbon dioxide and water are combined to produce carbohydrates (sugars) and oxygen. In most plants the leaf is the primary site of photosynthesis, which takes place in cells in the leaf interior. Photosynthesis depends on the absorption of light energy by plant pigments. Chlorophyll, the most important photosynthetic pigment, is located in oval cell structures called chloroplasts. Chlorophyll absorbs light throughout the visible range, although it absorbs green wavelengths poorly. (It is because chlorophyll reflects and transmits green wavelengths of light that most plant tissues appear green.)

Photosynthesis is the best-known function of leaves, but leaves play other roles as well. Pores (stomates) on the leaf surface allow water vapor to escape from the leaf interior into the atmosphere. This process, called transpiration, cools plant tissue and plays a crucial role in the uptake of water and minerals through roots. Stomates also allow carbon dioxide, which is required for photosynthesis, to enter leaves. This gas exchange (water vapor out, carbon dioxide in) takes place when stomates are open, but in times of excessive water loss, stomates may close to conserve water, thus reducing photosynthesis in the process.

Most leaves have a thin, waxy layer (the cuticle), which lies over a surface layer of nonphotosynthetic cells known as the epidermis. It is in the epidermis that the stomates are embedded. The cells in which photosynthesis takes place are between the upper and lower epidermis. In broad-leaf plants, stomates (plate 6) are usually more common on the lower leaf surface, where they may number in the hundreds of thousands per square inch (tens of thousands per square centimeter). In grasses and other plants having vertically oriented leaves, stomates tend to be numerous on both leaf surfaces.

Leaves have many other intriguing functions. They may attract ants and other insects with extrafloral nectaries, nectar-producing structures formed at the base of leaves. The benefit to plants from this arrangement comes when the ants attack would-be herbivores. Leaves also may draw excess water away from the stem, causing water to drip from their tips—enabling plants to dry off in wet habitats such as tropical rainforests—and in many species, leaves hinder small herbivores as they attempt to move about the plant to feed. Leaves also may provide the plant with supplemental nutrients, as in the carnivorous Venus flytrap (*Dionaea muscipula*; plate 7), pitcher plants (*Sarracenia* and *Nepenthes*), and sundews (*Drosera*).

Leaves provide food for the plants themselves, as well as for humans and other animals. Indeed, humans have discovered many uses for leaves. Leaves of sisal (*Agave sisalana*) and Manila hemp (*Musa textilis*) are harvested for their fibers, useful in making rope. Dried grass leaves traditionally have been used to thatch roofs. Maintenance of green-leaf grasses, especially Kentucky bluegrass (*Poa pratensis*) and fine-leaf fescues (*Festuca*), is at the heart of the multibillion dollar lawn-care industry. Cultivation and processing of leaves also provides the basis for a worldwide beverage industry; this is particularly true of the tea plant (*Camellia sinensis*, *Thea sinensis*) used in producing black, green, and oolong teas. Other leaves are used as herbs (for example, parsley, *Petroselinum sativum*) and drugs (tobacco, *Nicotiana tabacum*; cocaine, *Erythroxylon coca*).

Stems

The primary functions of aboveground stems are support of aerial plant parts and transport of water and nutrients. Stems help leaves reach the light needed for photosynthesis by providing physical support, elevating them above the ground, and positioning them in multiple tiers (plate 8). In shrubs and trees, woody stems allow plants to grow quite tall, placing leaves where they are less likely to be shaded by competitors. Stems also support flowers, fruits, and other structures; transport water and dissolved substances through conducting tissues; provide a means of vegetative spread; give protection from herbivores; serve as storage organs; and may themselves photosynthesize (plate 9).

Materials are transported through the stem primarily in two tissues. Water and dissolved minerals that enter the plant through the roots are transported through the xylem. Xylem cells die at maturity and form a hollow plumbing system throughout the plant. Outside the xylem is phloem, which is comprised of living cells that transport the sugar produced by photosynthesis from leaves to other parts of the plant. Additional air spaces and columns of cells conduct gases, resins, and other substances. In trees with thin bark, the phloem is located close to the trunk surface—this is why such trees can be easily killed by girdling, which severs these tissues and thereby cuts off the supply of sugar to the roots. Similar damage can occur on portions of woody plants that are rammed by lawnmowers and other equipment. The constriction or breakage of conduction tissue is why herbaceous plants that are bent often die rather than simply continue to grow horizontally.

Stems also help plants reach and colonize new sites by means of vegetative spread. Horizontal stems may be produced on or below the ground surface. Strawberries (*Fragaria*) and lamb's ears (*Stachys byzantina*) send runners across the surface (stolons); these runners then send down roots (plate 10). If the runner is later severed, the newly established plant is capable of life on its own. Unfortunately for gardeners, crabgrass (*Digitaria sanguinalis*) also spreads vigorously in this manner. Other plants produce horizontal stems below the soil surface (rhizomes). Roots are induced to form along the length of rhizomes, which leads to the production of new shoots. Irises and many ferns spread in this manner; this is also how quackgrass (*Agropyron repens*) and Johnson grass (*Sorghum halapense*) find their way into the garden. Some plants spread when a stem or branch touches the ground and is stimulated to send down roots; wild raspberries (*Rubus*) frequently do this.

Gardeners and breeders use stems to propagate plants. Woody plants such as currants and gooseberries (*Ribes*) and *Rhododendron* can be propagated by laying a stem or branch on the ground and covering it with soil (a practice called layering, or stooling). Many woody plants can also be propagated by wounding aerial branches and then wrapping the wound with moist peat moss until roots emerge from the stem into the moss. The section of stem with new roots can then be severed and planted elsewhere (a practice called air layering, or marcottage). Grafting is often used, although more by professional growers than by gardeners, to combine a branch or stem of one plant to an established root system of another. This is especially effective with woody species such as grapes and fruit trees.

Humans rely to a tremendous extent on the stems of woody and herbaceous plants. The stems, or trunks, of trees—in particular, the xylem—provide wood used worldwide for construction, cooking, heating, and the manufacture of paper and thousands of other products. Stems used as food include the white potato (*Solanum tuberosum*) and sugar cane (*Saccharum officinarum*); those used as spices include ginger (*Zingiber officinale*) and cinnamon (*Cinnamomum zeylanicum*). Stems also provide fibers, such as those from kenaf (*Hibiscus cannabinus*); gums, such as chicle (*Manilkara zapota*), the basis of chewing gum; resins, such as those obtained from various pines and used to make wood and rope resistant to saltwater; and in the case of the so-called diesel oil tree (*Copaifera langsdorfii*), a sap that can be used as a fuel oil. Medicines obtained from stems include quinine (*Cinchona*) and taxol, effective against certain cancers and originally extracted from the bark of the Pacific yew (*Taxus brevifolia*).

Roots

Aboveground plant parts, although more familiar than roots, depend greatly on their underground counterparts. Roots take up water and dissolved nutrients, anchor plants in place, store nutrients, and perform other specialized functions. In quaking aspen, they provide a means of vegetative spread; in the prop roots of corn/maize (*Zea mays*; plate 11) and the buttress roots of many tropical figs (*Ficus*), roots support the stem. The aboveground roots of red mangrove (*Rhizophora mangle*) absorb oxygen. In sweet potato (*Ipomoea batatas*) and cassava (or manioc, *Manihot esculenta*), roots store large amounts of carbohydrates; and in manroot (*Marah*), they store water.

Roots come in many forms, but a common and useful distinction is that between fibrous roots, which consist of many, usually thin, roots of approximately equal size, and taproots, which generally have one root that is significantly larger than the others. Fibrous roots are common in grasses, and they contribute to the thick, organic-rich soil characteristic of tallgrass prairie and other grasslands that receive adequate water. In many plants having fibrous roots, particularly grasses, the biomass of the root system is greater than that of the shoot system. Taproots, on the other hand, tend to grow deeper than fibrous roots, anchoring the plant and potentially reaching deeper water. Taproots are characteristic of woody trees and shrubs, although other plants have them as well. Some plants, including clover (*Trifolium*), have root systems that share features of both fibrous roots and taproots.

Roots do not always grow where and how we expect them to. Adventitious roots arise from the shoot system (plate 12). They form on underground stems such as fern rhizomes, from black raspberry (*Rubus occidentalis*) branch tips that come in contact with soil, along the stems of many vines, and as prop or buttress roots in corn/maize. Roots may also form at the base of cuttings placed in water, which is a common indoor means of propagating *Philodendron* and coleus (*Solenostemon scutellarioides*).

Roots increase in length as a result of cell division and expansion just behind the root tip (see section on meristems). As the root grows, its tip is pushed through the abrasive soil. Covering the root tip is a protective cap surrounded by a lubricant that helps the root thread its way between soil particles. Despite the presence of this cap and lubricant, the root tip is abraded and subject to damage. This is one of the reasons why most garden plants do best in a loose, well-aerated soil.

Figure 1. Although there are exceptions, dicots, including the dandelion (*Taraxacum officinale*, left), characteristically have a taproot. Monocots, such as this red fescue (*Festuca rubra*, right), typically have fibrous root systems.

Another intriguing aspect of root growth is obvious, but not entirely understood—roots grow down, whereas, for the most part, stems grow up. Major roots such as taproots are strongly influenced by the gravitational pull of the Earth. Smaller roots are generally less strongly influenced by gravity, enabling them to also grow sideways and upward. The root cap plays a central role in the root system's response to gravity, in part due to the presence of organelles in their cells that contain starch grains. These organelles, the amyloplasts, settle within cells in response to gravity, and this then affects root growth.

Roots can extend through a great soil volume and can reach great depths, but they take up relatively little water across most of their surface area. By far, most absorbed water enters roots through root hairs, which are tiny projections of epidermal cells that can number as many as 348,000 per square inch (40,000 per square centimeter). Root hairs are concentrated in a small area a short distance back from the root tip (plate 13). Despite being so tiny, root hairs absorb large amounts of water—provided, of course, that water is available in the soil. Root hairs are also delicate, however, and they are inevitably damaged when plants are repotted or transplanted. This damage results in reduced water uptake and explains why plants often fare poorly when moved, even if handled carefully. Thus, it is wise to reduce a plant's water needs at such times, for example, by shading or reducing leaf surface area through pruning. It also helps explain why garden plants do best if transplanted when water needs are at a minimum, such as in the evening or during damp weather.

Conducting tissue is located at the center of roots. Absorbed water, therefore, must be transported from root hairs at the epidermis to the root center. Once water reaches the plant's central xylem, it is drawn up into the stem by the forces of water cohesion generated as a result of transpiration as well as other forces generated by the roots. Roots also take in many substances that are dissolved in the water. Among these are dozens of nutrients and minerals required by plants (discussed in chapter 4), as well as compounds that are not needed and may even be toxic. The root does have mechanisms to exclude some molecules, which may, to a limited degree, protect garden plants from soil contaminants. Even so, it is best to place gardens away from sites likely to contain toxic substances, for example, roads and driveways where gasoline and oil are known to have been spilled.

Roots do not act alone in taking up water and minerals, but form interdependent partnerships with other organisms. Mycorrhizae are associations between roots and fungi in which the plant gains increased efficiency of water and nutrient uptake and the fungi gain access to nutrients supplied by the plants. Formation of these associations seems to be the rule among plants. More easily observed than mycorrhizae are root nodules, which result from associations between bacteria and a limited number of plants, particularly legumes (family Fabaceae or Leguminosae) and alders (*Alnus*). The bacteria contribute nitrogen to host plants and gain carbohydrates and physical protection in return. These two mutually beneficial associations are discussed in chapter 5.

Roots are as important to the health of garden plants as are other plant parts. In fact, plants can often survive total loss of aboveground tissues if sufficient food reserves are available in roots to enable regrowth—gardeners are all too familiar with the regrowth of weeds that have been chopped off at ground level. However, the gardener's goal is to promote healthy root growth by keeping aboveground tissues healthy; minimizing, to the extent possible, herbivores and pathogens (both above- and underground); and maintaining soil with appropriate physical and chemical characteristics in which plant roots will thrive.

Carrot (*Daucus carota* ssp. *sativus*), turnip (*Brassica rapa*, Rapa Group), and many other plants have nutritious roots and are widely consumed by humans, and the root of horseradish (*Armoracia rusticana*) is a common condiment. Roots are also a source of drugs, such as the intoxicating beverage kava (*Piper methysticum*), and insecticides, such as rotenone made from plants including cubé (*Lonchocarpus*). Prior to the advent of synthetic dyes, nearly all dyes were obtained from plants; there is renewed interest in natural plant pigments such as those extracted from bloodroot (*Sanguinaria canadensis*) and alkanet (*Anchusa officinalis*).

Flowers

Flowers come in a huge array of sizes, colors, and shapes. Some flowers are strikingly obvious, whereas others are so reduced and modified as to be nearly unrecognizable. Flowers may be purple, red, green, white, and every color in between; they may be spotted, striped, mottled, or variegated; they may be symmetrical or asymmetrical. Despite this variety, a flower's purpose—with very few exceptions—is to provide plants with a means of sexual reproduction (plate 14). Flowers are also the primary reason why gardeners grow many ornamental plants.

Flowers may form at the tip of the main stem, as in yellow bell (*Fritillaria pudica*) and shooting star (*Dodecatheon*). They may appear along the upper part of the stem, as in foxgloves (*Digitalis*) and penstemons (*Penstemon*). Flowers may tuck into the axils between leaves and stems, as in many mints (family Lamiaceae or Labiatae), or they may originate in clusters to form large, conspicuous arrays, as in dill (*Anethum graveolens*) and blanket flowers (*Gaillardia*)—to name just a few possibilities. These flower arrangements are important ecologically (for example, in their effects on pollinators) and affect the plants' desirability as garden plantings.

Most familiar flowers have four main parts arranged in concentric rings, or whorls: the sepals, petals, stamens, and carpels. The outermost of

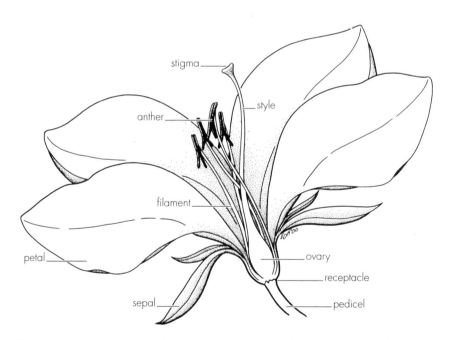

Figure 2. There is no such thing as a typical flower. But many familiar garden flowers, especially those pollinated by insects, include four main parts (from outside to inside): sepals, petals, stamens (each comprised of a terminal anther and a slender filament), and carpels (often comprised of a terminal stigma, a slender style, and a basal ovary). These flower parts originate from an expanded region, called the receptacle, at the end of the flower stalk, or pedicel.

these whorls is comprised of the sepals, which are often green and leaflike. (In fact, each of the main flower parts is derived from leaves, having been modified by natural selection in sometimes spectacular ways.) Sepals may also be other than green, and in these cases they are frequently similar in color, shape, and size to the petals, as in onion (*Allium cepa*) and magnolias. When this is true, the sepals often contribute to pollinator attraction. Before a flower opens, the petals and other flower parts are folded up tightly inside the protective sepals. Most flowers open slowly, but some, such as moonflower (*Ipomoea alba*), open so quickly that you can actually see it happening—it's time well spent sitting and watching a flower open and stretch as if waking from a cramped sleep. In horticultural varieties, sepals have frequently been modified by plant breeding, such as in ornamental irises (*Iris*), in which petals are "standards" and sepals are "falls."

Petals are tucked inside the flower's sepals. When petals are conspicuous and colorful, their chief function is nearly always to advertise to pollinators the presence of pollen and nectar. Although predictable pollinator–flower color pairings do exist, for example, hummingbirds and red flowers, these associations are now thought to be less rigid than once believed. Pollinators may favor a particular petal color or even a particular plant species, but circumstances such as how common a plant species is or the quantity or quality of pollen or nectar rewards may cause them to switch to plants with different-colored petals. As is true of sepals, petals can be independent or fused with each other or with other flower parts. In honeysuckle (*Lonicera*), the petals are largely fused with each other, forming a narrow cylindrical tube, at the base of which is a sweet drop of nectar (plate 15). This narrow tube, as in other flowers with similar shape, restricts which insects can reach the nectar, limiting access to species that are likely to transfer pollen effectively to the next honeysuckle flower. The petals of orchids (Orchidaceae) are especially renowned for their spectacular diversity. Orchid flowers typically have three petals, one of which differs from the other two and is often modified into a lip, platform, bucket, or other structure that reflects the anatomy and behavior of that orchid's pollinator.

Just inside the petals are the stamens, which are composed of a pollen-containing anther borne at the end of a thin stalk, or filament. A flower may have a single stamen, as in most orchids and many grasses, or many stamens, as in poppies (*Papaver*) and cherries (*Prunus avium, P.*

cerasus). Anthers open in a variety of ways to make pollen available. Pollen grains are tiny—at maturity they consist of only three cells. Despite their diminutive size, however, they are critical to the life cycle of flowering plants, because two of these three cells are the sperm cells necessary for fertilization. Pollen grains are often intricately sculpted and are highly resistant to decay.

At the flower's center are one or more carpels; if more than one, carpels may be separate or partly or wholly fused. (The term *pistil* is commonly used in referring to the carpels, whether separate or fused.) A typical carpel includes a terminal stigma, an expanded surface on which pollen grains are deposited. Holding the stigma aloft is a style, through which the pollen grain's sperm cells must be transported. At the base of each carpel is the ovary, in which the eggs are produced and the seeds (the product of eggs plus sperm) develop.

Although many flowers have the four-whorled structure just described, it is also common for one or more whorl to be absent. Wind-pollinated flowers frequently lack sepals, petals, or both, to minimize interference with transfer of wind-dispersed pollen. Other flowers have stamens but not carpels or vice versa. In cucumbers (*Cucumis sativus*), the stamen-bearing and carpel-bearing flowers are produced on the same plant, whereas in sweet gale (*Myrica gale*) and willows (*Salix*), male and female flowers are on different plants. The maidenhair tree (*Ginkgo biloba*) is an interesting example of a plant having separate male and female individuals. Many landscapers plant only the male trees because the females produce a "fruit" having a most unpleasant odor (*Ginkgo* is not a flowering plant, and it does not produce true flowers and fruits.) However, in eastern Asia, the "nut" from the female plants is highly valued for medicinal and culinary uses.

Growth and Development

Growth and development are fundamental processes of living organisms, including plants. As a growing plant obtains needed resources, it increases in size and adds parts such as roots, branches, and flowers. A major characteristic of growth in most familiar plants—in contrast to that in most familiar animals—is that it is indeterminate, meaning that as long as conditions are favorable, plant growth can continue. At the heart of growth and development is the production of new cells, followed by enlargement and differentiation of these cells to perform specialized functions. By understanding the key features of plant cells, it is easier to appreciate how plants do—or do not—succeed in natural and human-created environments, including gardens.

Cell Structure

The diversity of the plant kingdom is staggering and likely not yet fully appreciated. Nevertheless, every plant is comprised of cells that are surprisingly similar. A cell randomly selected from the leaf interior of a plant growing nearby would be remarkably similar to a cell randomly selected from another plant halfway around the globe—so similar, in fact, that even a botanist might not be able to tell you which plant the cell was taken from.

Although microscopic in size, a plant cell is a complex structure and the site of tens of thousands of chemical reactions. The outermost layer of a plant cell is the cell wall, a reinforced perimeter constructed primarily of cellulose, which is the most abundant polymer on the planet. Cellulose is formed from thousands of linked sugar molecules and is chemically similar to starch, although it is much less water-soluble (plate 16). Just inside the cell wall is the elastic cell membrane, whose functions include regulating the passage of substances into and out of the cell and the assembly of cellulose. When a plant dehydrates, the cell membrane pulls away from the cell wall, a trauma that, depending on circumstances, the cell may or may not survive.

Within the cell membrane is a fluid-filled compartment containing a variety of largely interconnected organelles (subcellular bodies that are structurally and functionally specialized). Although it may not be the cell's largest organelle, the nucleus is the command center of a living cell. The nucleus contains the cell's chromosomes, which are comprised primarily of DNA and whose genes control the chemical reactions and cellular activities that keep organisms alive. The number of chromosomes per cell is characteristic of particular species and varies widely. Chromosome number in plants ranges from four in the composite *Haplopappus gracilis* to well in excess of 1000 per cell in some ferns. Chromosome number is useful in tracing evolutionary ancestry. From a horticultural perspective, plants whose chromosome numbers have been increased through breeding are sometimes larger than their parents and often have larger, showier flowers.

Perhaps the hallmark of plant cells are the chloroplasts, organelles in green leaf and stem-surface tissues in which photosynthesis takes place.

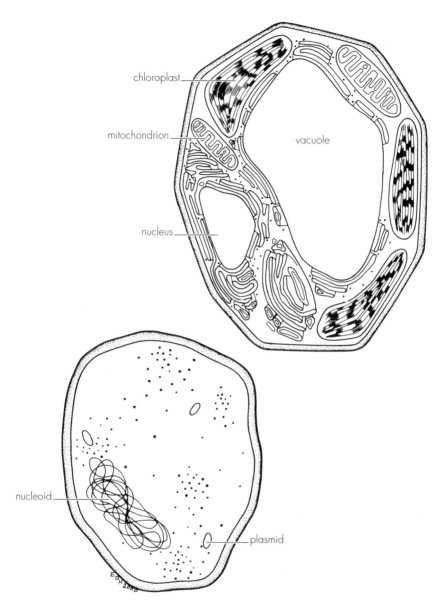

Figure 3. A mature plant cell (top) is usually dominated by the central vacuole. Other important organelles include the nucleus, chloroplasts, and mitochondria. A bacterial cell (bottom) lacks a nucleus, chloroplasts, mitochondria, and other organelles typical of plant and animal cells. Genetic material, usually in the form of a single, large chromosome, is often clustered into an area called the nucleoid. Some bacteria also have additional, much smaller pieces of DNA, picked up from the environment or other bacteria, called plasmids.

The green pigment chlorophyll, which is essential for the process, is contained within the chloroplasts (plate 17). Although photosynthesis is actually a rather complex series of reactions, these are generally summarized as carbon dioxide combining with water to yield the sugar glucose and oxygen gas. The energy driving these reactions is provided by light from the sun or from an alternate source, such as supplemental grow lights. Photosynthesis in most plants proceeds as described above, but there are alternative biochemical pathways, especially in plants that inhabit hot, dry environments.

The sugar produced in photosynthesis is the major energy source used by plants for metabolism, growth, and reproduction. This sugar may be used by the leaf in which it was synthesized, it may be metabolized in other plant tissues, or it may be stored in organs such as roots. Oxygen, which is a by-product of photosynthesis, may escape into the atmosphere or it may be used in chemical reactions within plant cells. For the energy of sugar to be used, its chemical bonds must be broken. This is accomplished in plants, as in animals, by cellular respiration, which occurs in cell organelles called mitochondria. In respiration, the chemical bonds of sugar are cleaved in the presence of oxygen to yield carbon dioxide, water, and energy. The chemical energy released then drives cell metabolism. Although the summarized chemical reaction for respiration suggests that this process is the exact reverse of photosynthesis, there are many alternate intermediate reactions, the details of which are beyond the scope of this book.

In living plant cells, chloroplasts tend to lie along the outer walls of the cells, whereas most of the cell volume is taken up by a vacuole, a fluid-filled sac that is bound by a membrane similar to that of the cell membrane. A cell's vacuole can occupy 90 percent or more of its volume. The vacuole, like an inflated water balloon, helps to maintain pressure within cells by virtue of the fluid it contains. It also stores many dissolved molecules, including proteins, toxins, and pigments such as anthocyanins, which impart red, blue, and purple colors to many flower and leaf tissues.

Many additional organelles and structures within cells perform critical functions—synthesis of proteins, nucleic acids, and other molecules; transport of materials within the cell; transport of materials toward and across the cell membrane; removal or recycling of wastes or damaged organelles; and much more.

Growth and development is all about cells: cell formation, cell growth, cell specialization, cell organization into tissues, and so on. As different types of cells are organized into various patterns, familiar plant organs such as leaves, stems, and roots are formed. These vegetative tissues, together with flowers, bring flowering plants to the stage at which, if all goes well, sexual reproduction may occur. If successful, a plant will produce seeds that bridge one generation to the next. The seed stage represents a convenient benchmark at which to begin a close look at the life of a plant.

Seeds and Fruits

The life of most seeds begins at the moment of fertilization of ovules by sperm, for it is at this point that an embryo begins development (see discussion of sexual reproduction later in this chapter). Once formed, seeds may require days, weeks, months, or even years to fully mature. A mature, viable seed includes an embryo and nutritive tissue within a protective seed coat. In some seeds, such as in peanuts (*Arachis hypogaea*), the nutritive tissue takes the form of cotyledons, or seed leaves (plate 18). The final product is packaged to withstand challenges of the environment that may include extreme temperatures, desiccation, abrasion from soil particles, passage through the digestive system of animals, fire, and other environmental insults. Seeds are so well packaged, in fact, that some have been known to survive in a dormant state for hundreds of years before successfully germinating. In the remarkable case of the sacred lotus (*Nelumbo nucifera*), a seed radio-carbon dated to be 1300 years old was successfully germinated. (Some earlier, less-rigorous studies claimed germination of lotus seeds more than 2000 years old.)

As seeds develop, tissues of the ovary and often of other flower parts differentiate and proliferate to form a fruit, which characteristically forms three layers around the seeds (plate 19). These layers may be fused or difficult to distinguish or they may be quite distinct. A classic example is the peach, with its thick, hard endocarp that protects the seeds; the fleshy fruit; and its outer skin. The inner of the three fruit layers may be papery, as in apples (*Malus*), or soft, as in grapes. In beans (*Phaseolus*), the outermost layer becomes hard and dry to form a pod. In grains, the fruit layers around the seed are so fused and reduced as to be indistinguishable, whereas in maples, these layers form the wing that causes the fruit to rotate as it drops toward the ground.

Fruit maturation usually requires fertilization and seed development, but the cultivated banana (*Musa ×paradisiaca*) develops without fertilization. In other cases, such as in seedless grapes, artificial hormones may be applied to flowers to induce formation of fruits without fertilization or to synchronize fruit production throughout a crop, making harvest more efficient.

Fruits, like the seeds they contain, are dispersed primarily by wind, water, gravity, and animals. Common wind-borne fruits include the parachute-like plumes of dandelion and the samaras of maples, whereas the coconut palm (*Cocos nucifera*) is perhaps the best-known water-dispersed fruit. Gravity may be a primary means of dispersal, as when fruit drops into the mud at the base of a parent plant, or it may act in concert with wind and water. Animals—including in some cases humans—can transport fruits great distances in their fur or hair, in mud on their bodies, and in their digestive systems. Fruits having barbs, hooks, and other means of attachment include those produced by bedstraw (*Galium*), beggar's ticks (*Bidens*), cocklebur (*Xanthium*), and tick-trefoil (*Desmodium*). Especially in dry fruits, the seeds within may be spilled as the fruit is transported. In other cases, a fruit may be carried off by an animal to a new location; the fruit may then rot or be consumed, with the seeds being deposited in the process. Gardeners who wish to save seeds must harvest fruits before they are dispersed by the wind, pecked open by a bird, consumed by an invertebrate, or fall to the ground.

Fruits and seeds have tremendous economic importance, particularly in the form of the grains, such as rice (*Oryza sativa*) and wheat (*Triticum*), which sustain Earth's human population. A list of other seeds and fruits used as foods and beverages would be nearly endless, but could be started by surveying the shelves of a local grocery store. Other uses of seeds and fruits as herbs, spices, and medicines are plentiful.

Seed Germination

Eventually, a seed will leave its dormant state and germinate or it will die. Seed germination is a tricky business, and, as in doubles figure skating and trapeze work, timing is crucial. In temperate climates, seeds that germinate in the spring must do so early enough to take full advantage of the growing season and get a jump on competitors, but not so early that the young seedlings are killed by a late frost or by cold-loving fungal pathogens. Seeds frequently do not get the timing right, and many seedlings perish before reaching maturity.

Numerous mechanisms increase the likelihood that seeds will germinate at an appropriate time. Seeds of many desert annuals, for instance, will not germinate until an inhibitory chemical such as a phenolic has been sufficiently leached from the seed. This happens only when seeds have absorbed adequate amounts of soil water following rain—the perfect time for a desert seed to germinate. Seeds growing in areas with cold winters usually must be exposed to sufficiently low temperatures for a minimum period of time to germinate (stratification). Seeds of ginseng (*Panax*), coneflower (*Echinacea*), and many other temperate-zone plants must be planted in autumn and allowed to pass through a winter (or kept in wet sand or some other medium in a refrigerator for weeks or months) before they will germinate. Serviceberry (*Amelanchier*) seeds require cold stratification if the fresh ripe berries are used as seed, but a double stratification of cold-hot-cold (mimicking two winters) if the harvested seeds are allowed to dry out. Seeds of birch (*Betula*), bell flower (*Campanula*), and some varieties of lettuce (*Lactuca sativa*) contain a chemical inhibitor that breaks down when exposed to light; only then will these seeds germinate. More specifically, these seeds germinate in response to the red wavelengths in sunlight. Seeds that are light-sensitive are usually small; by germinating only when exposed to at least a minimal intensity of light, seedlings are more likely to establish successfully. Were such small-seeded plants to germinate deep within the soil, they likely would not have adequate reserves to reach the surface and begin photosynthesizing. Geranium (*Geranium*) and California poppy (*Eschscholzia californica*) seeds, in contrast, germinate only in the dark. Still other seeds, including those of American holly (*Ilex opaca*) and many orchids, contain an immature embryo at the time they are released from the maternal plant. These require additional time to complete maturation, a process called after-ripening. Seeds of plants such as avocado (*Persea americana*), however, germinate best while still surrounded by fresh fruit tissues, and some gardeners plant the entire fruit.

Some species must have their seed coats physically abraded or cracked (that is, they must undergo scarification) to germinate. This can happen when a seed is blown across a rough soil surface, subjected to alternate freezing and thawing, or passed through the digestive system of an animal. Gardeners who do not keep a menagerie on hand for this purpose sometimes accomplish the same end by rolling seeds in a mechanical tumbler, thinning the seed coat with sandpaper, or by nicking it with a sharp file or razor blade—being careful not to damage the embryo, of course.

Most seeds purchased from garden centers and through catalogs either come from plant varieties that have had specialized germination requirements bred out of them or that have already been subjected to required preconditioning by the time of purchase. Special germination requirements are more likely in seeds collected from the garden or those of heirloom or undomesticated wild plants. Gardeners who save seeds from year to year should also pay close attention to storage conditions, as most seeds maintain viability longest if kept cool and dry, although seeds of many tropical plants do not.

Lest we think we are especially clever, there are many seeds whose germination secrets we simply haven't deciphered. For years, Steve Carroll and his students have tried, unsuccessfully, to germinate the seeds of bird's foot violet (*Viola pedata*). Attempts so far have included planting seeds indoors and outside at different times of the year and in different types of soil, storing the seeds for varying lengths of time at different temperatures, sanding and nicking the seed coat, soaking the seeds in sulfuric acid, exposing the seeds to varying concentrations of gibberellic acid and carbon dioxide and various day lengths, and ordering the seeds in a loud, firm voice to "Germinate!" Still, the seeds refuse to cooperate.

Seedling Emergence

Many challenges lie ahead for a seed that manages to germinate. The primary root (radicle) is usually first to emerge through the seed coat of a germinating seed; this root must penetrate the soil and begin taking up water and nutrients. The primary shoot grows in the opposite direction to reach light, and it must begin photosynthesizing before the seed's energy reserves run out.

In many plants, the energy required for this early growth is provided by the cotyledons. In kidney beans (*Phaseolus vulgaris*) and peanuts, for example, most of the seed's volume is taken up by the two cotyledons. As young bean plants begin to grow, the elongating shoot is bent into a hook shape. As a result the delicate shoot tip and cotyledons are pulled, not pushed, through the abrasive soil. Once the shoot reaches light, the stem straightens out, chlorophyll is produced, the cotyledons turn green, and photosynthesis begins. In contrast, in peas (*Pisum sativum*) and oaks (*Quercus*) the cotyledons and the seed coat remain below the soil surface. Seedling emergence in grasses is different still. Their shoots grow straight up through the soil inside a protective sheath. Simultaneously, the root

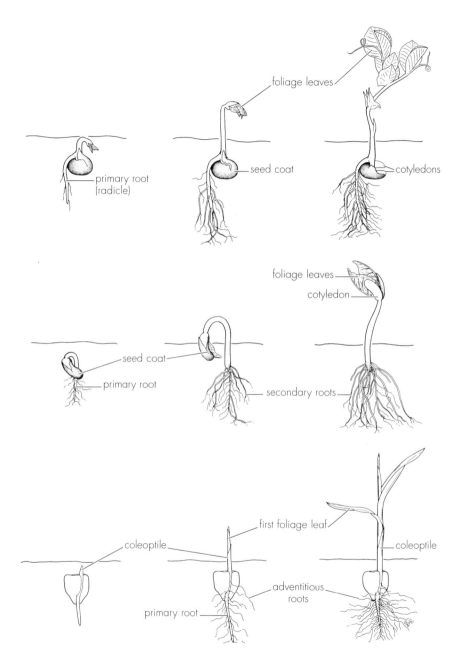

Figure 4. Seed germination and seedling growth follow characteristic patterns among dicots in which the cotyledons remain at or below the soil surface (represented by pea; top), dicots in which the cotyledons are pulled above the soil surface (represented by bean; middle), and monocots (represented by corn; bottom).

grows downward, also inside a sheath. Eventually, the shoot and root break through their sheaths and continue to grow without need of this protection.

Growth

Plants grow through two processes: production of new cells and increase in cell size. These two processes sometimes follow each other closely and sometimes do not. They are controlled independently, however, and both are necessary if a plant is to exhibit normal, healthy growth.

New cells form from the division of existing cells in tissue known as meristems. When a meristem cell divides, one of the daughter cells is able to divide again, whereas the other develops into a more specialized cell. Because meristems continue to produce new cells, a plant is able to continue growing throughout its life.

Meristems are located in characteristic places, depending on the type of plant. They form at stem and branch tips in dicots (broad-leaf plants; plate 20). Division of cells in these meristems causes increased plant height and branch length. Because broad-leaf stems grow from the tip, rather than from the base, branches are at much the same height above the ground years after emerging. Growth also occurs from terminal meristems in roots, allowing them to reach sometimes surprising depths and distances. In monocots, such as grasses, meristems are located at nodes along the stem, as well as at the apex. As cells in these meristems divide, new tissue is formed toward, rather than away from, the tip of the stem. As a result, when an herbivore—or a lawn mower—removes the tip of a grass plant, the plant is able to continue vertical growth from meristems lower on the shoot (plate 21). Broad-leaf weeds, on the other hand, may have their meristems destroyed by mowing, which is why mowing may benefit a lawn—by favoring grasses over broad-leaf plants.

Plants have meristems at other characteristic locations, such as in buds in the axils of branches and leaves. Although these meristems are often inactive, their cells may start dividing when a plant's main shoot tip is eaten by a rabbit or trimmed by a gardener. This bushing out of a plant may or may not be desirable from a gardener's point of view. Woody plants characteristically have a cylindrical meristem below the stem surface. Division of cells in this meristem leads to an increase in diameter, rather than an increase in height. In habitats in which growth ceases for part of the year, seasonal activity of this meristem forms the annual rings characteristic of many woody species.

After new cells have formed through cell division, they may begin enlarging. Cells increase in size by adding materials to their cell walls and membranes. By adding more materials in certain dimensions, a great variety of cell shapes can be achieved. In addition to increasing in size, plant cells also differentiate as they assume specialized functions. In flowering plants, for instance, cells that differentiate as fibers (to provide strength) are characteristically longer and thicker-walled than those that conduct the sugars produced in photosynthesis. Cell enlargement also depends to a great extent on internal pressure generated by the vacuole. For a plant to provision enlarging cells with water, there must be adequate water available in the environment. Therefore, growth, like so many other plant processes, is intimately linked to the supply of water.

Plant Growth Form

Meristematic growth has many consequences for the overall form of plants, which greatly influences which habitats a plant can grow in. For example, a plant having weak apical dominance and many axillary meristems—therefore producing a highly branched, bushy growth form—would have a large surface area and would likely do poorly in an arid, windy environment, but might be well suited to a moist forest floor.

Plants tend to continue growing as long as conditions remain suitable, and most do not have an immutable shape, as evidenced by the growth patterns that result when parts of plants are damaged, diseased, or pruned. Understanding growth mechanisms and the locations of meristems in various types of plants is useful in the garden and home, as it allows gardeners to encourage a leggy plant to expand more fully or to change its shape to fill an available space. It is also at the heart of the practice of bonsai, in which woody plants are trained to assume pleasing forms.

Most garden and lawn plants, major crops, trees, and shrubs are rooted in the soil and are essentially self-supporting. Aquatics live most or all of their lives in water and may be free-floating or rooted in the bottom sediments. Vines begin life in the soil; however, they eventually come into contact with, then climb up, a tree, trellis, or another suitable vertical structure. In so doing, they lift their leaves above competitors to take advantage of better access to light. Vines include woody species such as trumpet creeper (*Campsis radicans*) and herbaceous species such as morning glory (most commonly, *Ipomoea purpurea* or *I. tricolor*). Epiphytes forego the ground altogether; their seeds germinate in crevices on the trunks and branches of other plants, and they obtain water and nourishment from rain and dissolved nutrients. Plants that grow in this way include most bromeliads and many orchids and ferns.

The growth form of an individual plant is greatly influenced by the environmental conditions in which it grows. For example, two plants of the same species may develop markedly different leaves depending on their light environment, with plants grown under low-light conditions having larger leaves than those grown under high-light conditions. Leaf form can even vary dramatically within an individual plant. In the aquatic, ribbon-leaf pondweed (*Potamogeton epihydrus*), submerged leaves lack a petiole and are linear, whereas floating leaves have a petiole and are oval and shorter. Many other aquatics show similar leaf dimorphism. Plant growth form can also be affected by the biological environment. Herbivory, for example, can increase the density of defensive structures such as spines and can alter plant shape (see discussion in chapter 5).

Plant Adulthood

The term *adult* is rather arbitrary when applied to plants, but it can be used to refer to the growth stage at which they are capable of flowering. Different plants reach this landmark on different timetables. Annuals—such as *Petunia* ×*hybrida* and most cereal grains—flower, produce seeds, and die during their first season of growth. In northern climates, familiar garden plants such as begonias (*Begonia*) and eggplant or aubergine (*Solanum melongena*) grow as annuals, although they live longer than one season in their native tropical habitats. Winter annuals, such as cheatgrass (*Bromus tectorum*) and chickweed (*Stellaria media*), germinate in late summer or autumn, grow, then flower and die early the following growing season. Biennials, such as carrots, grow during their first season, then flower, produce seeds, and die in their second.

Perennials generally live for many years, although they may not flower every year. Exactly when a perennial plant flowers and how frequently it does so depends on genetic variety, soil conditions, temperature, water availability, effects of plant pests, and other factors. Some carefully tended garden perennials may flower during their first growing season, whereas others, particularly woody species, may not flower for several years after planting. One variation on the perennial theme is found in century plants (*Agave*), some species of which grow for decades (although not necessarily a hundred years). They then flower once in a spectacular display and die. Some bamboos (*Bambusa* and *Phyllostachys*) are particularly intriguing, as they may persist for several decades in a vegetative (that is, nonflowering) state. Then, for reasons that are not clear, clones of an original mother plant flower simultaneously—regardless of how geographically separated they may be—set seed, and die.

Hormones

Plant growth and development may be easy to observe, but their underlying causes are not. Why does a plant suddenly undergo a growth spurt, send up a flowering stalk, or grow from a lateral branch? Part of the answer lies in a small number of organic substances called hormones. Despite their being present in tiny concentrations—sometimes as little as a few parts per million—hormones have a profound effect on plants. Hormones are produced in one part of an organism and transported to another part of that organism, where they have specific effects on growth and development that depend on, for example, their concentration, the tissue in which they are located, and whether they are acting alone or in combination with other hormones.

Charles Darwin and his son Francis were among the first to investigate the effects of plant hormones. They were interested in why plants grow toward light (a tendency known as phototropism). The Darwins devised clever experiments involving grass seedlings in which they either left the seedling tips uncovered or covered them with metal foil or transparent glass. They then compared growth of the seedlings in relation to a unidirectional light source, demonstrating that, in grass plants, the shoot tip played a role in causing growth toward light.

Modern botanists have discovered that one cause of such growth is presence of the hormone auxin. It stimulates increased cell elongation on the more shaded side of a stem, causing its tip to bend toward a directional light source. Those who grow indoor plants near windows often see this type of growth, which can be compensated for by periodically rotating plants. Auxin released from the shoot apex also inhibits growth from lateral buds. When the shoot tip is removed, either by pruning or by an herbivore, the lateral buds are released from this inhibition and growth becomes fuller.

Growers use auxin to stimulate root production in plant cuttings and to prevent premature dropping of apples and oranges (*Citrus sinensis*). Synthetic auxins are also used as herbicides to selectively kill broad-leaf plants, thus minimizing competition with grasses and other monocots.

Discovery of the plant hormone gibberellic acid, or gibberellin, arose from studies of rice plants (*Oryza sativa*) suffering from *bakanae*, or "foolish seedling disease." These plants, infected with the fungus *Gibberella fujikuroi*, were unusually pale and spindly, and they eventually fell over. Studies eventually showed that the fungus secreted gibberellic acid, which caused this abnormal growth. Later it was found that substances resembling gibberellic acid (gibberellins) are also produced by plants themselves. Gibberellins, of which there are dozens of variants, stimulate stem elongation, seed germination, and fruit tissue growth. Gardeners and commercial growers use synthetic gibberellins to break dormancy of hard-to-germinate seeds and also to facilitate production of seedless grapes.

The plant hormone abscisic acid (ABA) delays seed germination, retards growth, causes loss of leaves (abscission—hence, its name), and contributes to bud dormancy. Abscisic acid has relatively few practical applications, but because it causes stomates to close, it has been sprayed on landscape plants to decrease water loss during dry periods.

In flowering plants, hormones known as cytokinins are synthesized primarily in roots. These hormones, of which there are many chemical variants, stimulate cell division and influence the relative abundance of roots versus shoots. Synthetic cytokinins are sometimes used in tissue culture to encourage root growth in turf grasses and to keep cut flowers fresh. Also, cytokinins appear to be the active ingredients in seaweed extracts, widely used by organic gardeners to enhance plant growth and resistance to disease. Cytokinins also play a role in causing plant disease. Some insects, bacteria, and fungi produce cytokinins or cause plants to do so. This increased concentration of cytokinins, sometimes in combination with auxins, causes formation of galls and witch's brooms in olives (*Olea europaea*), oaks, and other species too numerous to mention (plate 22). In cases involving insects, these galls provide a means of protection and a source of nutrition for developing larvae.

Ethylene is the only currently recognized plant hormone that is a gas. It affects flowering, loss of leaves, stem elongation, fruit ripening, and many other processes. Ethylene produced by one plant can also affect nearby plants. Growers use this to advantage by storing unripe fruit, including grapes and tomatoes (*Lycopersicon lycopersicum*, *L. esculentum*), in rooms in which the air lacks ethylene. When they wish to market their fruit, they add ethylene to the atmosphere to induce ripening. This also explains why placing a ripe ethylene-producing peach (*Prunus persica*, *Amygdalus persica*) or apple in a paper bag containing unripened fruit causes the latter to ripen more quickly—a trick your grandmother may have taught you. Tossing a few overripe apples or bananas among tomato plants laden with reluctant-to-ripen green fruit can also accelerate their ripening.

Plant hormones affect virtually all aspects of plant growth and development, and every plant hormone has complex and multiple actions, including ones that seem contradictory at times. It is likely that new hormones await discovery and that known compounds will be elevated to hormone status. There is one particularly elusive hormone, widely called florigen, that may be at least partly responsible for inducing flowering. Although there is some evidence for this hormone's existence, the evidence is contradictory. As more is learned about plant hormones, the number of practical applications for these compounds will surely increase.

Sexual Reproduction

Sexual reproduction is one of nature's most remarkable events, and this is no less true in the plant world than in the animal world. We consider two stages in the unfolding of plant reproduction—pollination and fertilization. Pollination is a necessary act that enables the sperm and egg cells to find each other. During fertilization, genes from two individuals (unless a plant is capable of self-fertilization) combine to produce a new plant with a unique assortment of characters. In this way, sexual reproduction results in offspring, at least some of which are likely to have beneficial combinations of genes. Vegetative propagation (cloning), in contrast, yields offspring that are genetically identical to the parent plant. Although these offspring may be well suited to current conditions, it is thought that sexual reproduction confers advantages in the event that environmental conditions should change—which they inevitably will.

Pollination

Individual pollen grains must be magnified to be seen. But when present in large numbers, whether spread in a thin layer on the surface of a pond, in a film on a table top near an open window, or on the legs of a foraging bumble bee, pollen can be strikingly obvious (plates 23, 24). The term *pollination* strictly refers to the transfer of pollen from anther to stigma. This may be accomplished by wind, water, animals, or even without an outside agent (see below). Pollination varies with plant type, habitat, climate, and other factors. For instance, grasses and members of the beet family (Chenopodiaceae) are characteristically wind-pollinated. Water pollination is relatively rare, but is known in the aquarium plant ribbon weed (*Vallisneria spiralis*) and in species of waterweed (*Elodea*). Orchids are pollinated mostly by small bees, whereas night-blooming species, including stock (*Matthiola incana*), yucca (*Yucca*), and many cacti, are generally pollinated by moths and bats.

Pollination is a rather tenuous affair. In cold or wet weather, insect pollinators may be inactive, thereby reducing fruit and seed set. Similarly, honey bees (*Apis mellifera*) afflicted with infectious disease or infested by the increasingly problematic varroa mite (*Varroa*) will become less effective as pollinators—especially if they die. Wind-pollinated plants may receive insufficient pollen if there are not compatible pollen-shedding plants upwind. This is why corn/maize should be planted in blocks rather than in long, isolated rows.

Self-fertilization, or selfing, occurs when pollen is transferred from an anther to a stigma that is part of the same flower or is in a different flower on the same plant. Cross-fertilization, or outcrossing, occurs when pollen travels between flowers on different plants. Selfing has both advantages and disadvantages over outcrossing. Because pollen travels over shorter distances in selfing, pollination is generally simpler to accomplish, and because less pollen is likely to be wasted in the process, less need be produced. This increased efficiency can also yield greater numbers of seeds in situations in which pollen quantity is limited or cross-pollination problematic. However, selfing usually produces offspring with less genetic variation and a greater likelihood of genetic defects than offspring produced through outcrossing.

Despite these potential problems, selfing routinely occurs in plants such as marigolds (*Tagetes*) and *Dahlia*. An extreme example of selfing is seen in touch-me-nots (*Impatiens*) and many violets (*Viola*), which, in addition to their familiar, attractive, insect-pollinated flowers, also produce small, inconspicuous flowers. Because these small flowers do not open, pollen is automatically transferred from anther to stigma within the same flower. If the plant has available energy and resources, seeds may be produced through selfing. In this case, selfing acts as a backup system.

Other species balance outcrossing and selfing in different ways. *Hibiscus* anthers form a ring partway up a central column, at the apex of which are five stigmas supported by five short, robust styles. In the roadside weed flower-of-an-hour (*H. trionum*), which Steve Carroll and his students have studied, flowers open in the early morning. At that time, pollen is available and the stigmas are receptive. If a stigma receives pollen from a visiting insect, the styles remain relatively firm and the petals soon close, never to open again. But if the morning passes and little or no pollen has been received, the styles relax, causing the stigmas to bend forward and down until they come into contact with the pollen-containing anthers. Pollen that is still in the anthers can then be transferred to the flower's own stigmas, producing seeds through selfing. Although these seeds do not weigh as much as those produced through outcrossing, they result from a remarkable system that ensures production of at least some seed.

Wind-pollinated plants do not have "hired hands" ready to deliver pollen. Even so, flowers and inflorescences in these species have adaptations that increase the likelihood of successfully dispersing and receiving pollen. For example, structures in inflorescences of some wind-pollinated plants inhibit pollen release when conditions are calm; anthers in grasses usually dangle free of other flower parts, to increase exposure to the wind; and stigmas are frequently large, feathery, and exposed. Pollen produced by wind-pollinated species also tends to be smooth and dry; pine pollen even has air sacs that increase buoyancy (plate 25).

Considering all the places that pollen grains *can* end up—on the ground, in your nasal passages—it's a wonder that fertilization in plants is ever successful. But because pollen transfer is usually not random, at least some of a plant's pollen is likely to be delivered to the right kind of flower, to the correct part of that flower, and at a time when that flower is physiologically and anatomically ready to accept pollen.

Fertilization

It is no simple matter for a plant sperm cell to reach an egg. For fertilization to take place, sperm must be delivered from pollen grains on the surface of a stigma to egg cells in the ovary, which may be several inches away. Sperm cells in flowering plants do not have the whiplike tails of many animal sperm, so plant sperm cannot reach eggs under their own power. For fertilization to occur, pollen grains must first germinate on the surface of a stigma and then produce a pollen tube that penetrates the stigma, grows through the style, and reaches a small opening in the ovule, which develops into the seed following fertilization (plate 26). A pollen tube may grow 20 inches (50 cm) or more in corn/maize and even farther in other species. Once the pollen tube reaches and enters the ovule, it releases its two sperm.

Fertilization in flowering plants involves both of the sperm from the pollen grain, and for this reason is called double fertilization. One sperm cell combines with an egg to form the zygote, the first cell of the embryo, and the other sperm combines with other cells within the ovule to produce tissue (endosperm) that will nourish the developing embryo; the endosperm represents the bulk of the seed in the grains we eat, as well as in many other seeds.

Reproduction is often the most energy- and nutrient-demanding process in plant life cycles. Successful reproduction can be quite sensitive to such things as extreme temperatures or nutrient deficiencies. Gardeners who raise plants for seeds or fruits need to take especially good care of their plants during pollination and seed development.

Plant Life Cycles

A plant that reproduces sexually undergoes two major transitions in its life cycle. One occurs when particular cells in the anther and in the ovule divide to produce cells with half the chromosome number of their progenitor cells, a process called meiosis. In the anther, these resulting cells divide further—without a change in chromosome number—to form the pollen grain, two of whose cells are the sperm. In the ovary, corresponding cell divisions lead to the formation of individual egg cells within protective, multicellular ovules. The other transition occurs when sperm and egg cells unite at fertilization to form the zygote. In this transition, the chromosome number doubles, because those in the sperm and egg are combined, thus bringing the cycle full circle.

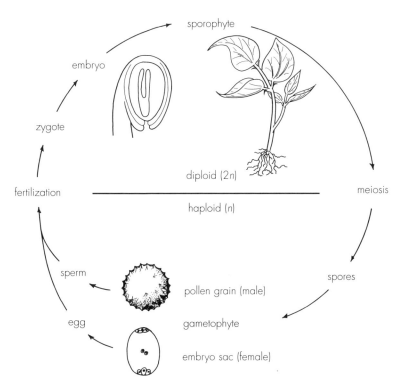

Figure 5. The life cycle of flowering plants involves two alternating phases (or generations), which differ by the number of chromosomes present. One phase of the life cycle is dominated by the diploid (2*n*), reproductive adult form of the plant, with its characteristic leaves, stems, and flowers (top). The other phase is entered when haploid (*n*) cells are formed through meiosis (bottom). The cycle is completed when a sperm and egg combine in fertilization to form the adult phase once again.

The two life stages that result from these transitions are called generations, which refers not to parents and their offspring but to phases in which chromosome number is alternately halved and doubled. One of the chief distinguishing characteristics of the plant kingdom is the alternation of multicellular generations. In flowering plants, the familiar tissues and organs that develop following fertilization (for example, seedlings, leaves) are comprised of cells containing two sets of chromosomes, which are therefore said to be diploid. The other generation is represented by the pollen grain and the embryo sac, each of which is multicellular, but whose cells are haploid, that is, they contain one set of chromosomes.

During the alternation of generations, errors sometimes occur. For example, if chromosome number is not appropriately halved, subsequent fertilization can result in an embryo having extra chromosomes. These types of changes in chromosome number have occurred frequently during the evolutionary history of plants. Doublings, and often subsequent redoublings, are thought to have occurred in most ferns and in as many as half of flowering plants. Unlike in animals, however, many agricultural and horticultural plants have been purposefully bred in this way. For example, common bread wheat (*Triticum aestivum*) has six sets of chromosomes (hexaploid), resulting from a cross between an ancestral species having four sets of chromosomes (tetraploid) and one that had two sets (diploid). There are several reasons why plant breeders intentionally double the number of chromosomes in plants. Sometimes it is necessary in order for two plants to be hybridized—a result of the way chromosomes pair during meiosis. In other cases, adding sets of chromosomes through doubling (or more) can increase plant or flower size, increase the seed or fruit crop, or lead to some other desirable characteristic.

Classifying and Naming Plants

For two people to discuss a particular plant—whether to order a shrub at a nursery or to discuss an interesting garden flower—the two must be speaking about the same plant. This is why systematics is so important. In general, systematics accomplishes two important tasks: organizing species into a logical, meaningful classification system and assigning appropriate names.

Nearly all organisms are classified according to a system with seven hierarchical categories: kingdom, phylum (sometimes called division in the cases of plants, fungi, and bacteria), class, order, family, genus, and species. These categories can also be subdivided into, for example, subphylum and subspecies, and often are in plant taxonomy. Using this hierarchy, species as different from each other as a human and the common lousewort can be conveniently classified with other closely related species (table 1). This system is not perfect—there is disagreement concerning the relatedness of different families, the number of species in particular genera, and more. By and large, however, this classification has stood the test of time, and today it serves as the foundation for a means of communication that links gardeners and other growers worldwide.

Table 1. Hierarchical classification of two organisms

Level	Human	Common lousewort
Kingdom	Animalia	Plantae
Phylum	Chordata	Magnoliophyta
Class	Mammalia	Magnoliopsida
Order	Primate	Lamiales
Family	Hominidae	Scrophulariaceae
Genus	*Homo*	*Pedicularis*
Species	*Homo sapiens*	*Pedicularis canadensis*

Without a logical classification system, gardeners would be faced with a long, unorganized, and overwhelming list of names. But there are countless ways that plants—and other organisms—could be organized. Flowering plants could be grouped by flower color, size and growth form, or the number of their flower parts. In fact, each of these characters has been used: field guides often organize plants by flower color, size, or growth form (for instance, herbs, shrubs, or trees), and Linnaeus, an eighteenth-century naturalist, organized plants based on their number of stamens. The classification system used today is based, to the extent possible, on hypothesized evolutionary relationships. In this system, species that are believed to be most closely related based on all available evidence are grouped together (plate 27). Evolutionary relatedness can sometimes be difficult to determine, but as genetic screening techniques are developed and improved, new information is obtained, and relationships are clarified, plants are regrouped to reflect this information.

Using common, rather than scientific, names for organisms can lead to confusion. If a gardener told two friends about a recently planted primrose, one friend might think it was a species in the genus *Primula*, a member of the primrose family (Primulaceae), whereas the other might think it was *Oenothera*, a very different plant in the evening primrose family (Onagraceae; plate 28). Using a common name is fine when standing in front of a plant in a particular garden, but many plants have dozens of common names that vary from one location, tradition, or language to another and others don't even have common names.

To avoid confusion and to create an internationally useful system for naming plants, a formal set of guidelines has been established. These are compiled in the *International Code of Botanical Nomenclature*, which requires that each species be assigned a unique, two-part Latin name. To the extent possible, names are assigned so that plants understood to be closely related are grouped together. For example, all dogwoods are placed in the genus *Cornus*. Such a scheme places related plants into a meaningful framework based on the best available botanical and evolutionary evidence.

Asclepias syriaca is the official botanical name for one of the plants commonly called milkweed. Taken alone, neither of the two parts of this name unambiguously refers to this plant. The first part of the name, *Asclepias*, is the genus in which this milkweed is classified, along with other related milkweeds (also called butterfly weeds). The second part of the name, the specific epithet, specifies which one of these related plants is intended—*Asclepias syriaca* and not *Asclepias tuberosa*, for example. By convention, genus and species names are italicized and the Latin name of a genus is always capitalized, whereas the epithet is not. A complete scientific name also indicates who first formally described the species or placed it in its genus. For example, the complete name for the common milkweed is *Asclepias syriaca* L., in which the letter "L." is the standard abbreviation for Linnaeus, who named this plant. The name associated with the yellow stemless primrose of the American Southwest, *Oenothera brachycarpa* A. Gray, refers to the nineteenth-century American botanist Asa Gray. Although many gardening books and plant catalogs omit these abbreviations, more advanced publications often include them.

The term *cultivar* (from "cultivated variety") refers to varieties of cultivated plants that have been produced through horticultural means. Horticultural species often include many different cultivars, and the gardener who orders a plant without specifying which variety he or she wishes is apt to be surprised. To avoid this, names for horticultural varieties are written in one of two ways, either by using single quotation marks (*Asclepias incarnata* 'Ice Ballet') or by using "cv.," an abbreviation for cultivar (*A. incarnata* cv. Ice Ballet). The names of cultivars, which are capitalized but not italicized, may also be used without the specific epithet, as in *Narcissus* 'Aflame'.

Gardeners are also likely to encounter several ways of naming horticultural hybrids. For example, hybrids are sometimes given their own names, as when a cross between *Aster* and *Solidago* is called ×*Solidaster* or when a cross between *Verbascum lychnitis* (white mullein) and *V. nigrum* (dark mullein) (*Verbascum lychnitis* × *nigrum*) is called *Verbascum* ×*schiedeanum*. When known, the seed parent should be listed first in a cross; otherwise, the parent plants should be ordered alphabetically.

☙❧

Although there are exceptions—for example, water gardens where the goldfish (*Carassius auratus*) and dragonflies are more evident than the plantings—plants are the most prominent life forms in most gardens. Nonetheless, anyone who grows plants knows that gardens are also populated by a never-ending parade of organisms ranging from itinerant flocks of voracious finches to web-building spiders to a seemingly endless list of microscopic and near-microscopic bacteria, mites, nematodes, and other organisms rarely or never glimpsed. In chapter 2, we examine some of these other organisms, especially in light of the ecological roles they play in gardens

2

Other Garden Inhabitants

Plants don't live alone in the garden. Depending on local conditions, plants and gardeners may share the garden environment with creatures ranging from elephants to viruses. These organisms can, and often do, exert powerful influences on the health and well-being of garden plants. Wild elephants (*Loxodonta*) may trample and devour, whereas domesticated elephants serve as capable beasts of burden. Raccoons (*Procyon lotor*) will travel miles to eat sweet corn, but also consume harmful insects and mollusks. On the one hand, insects may utterly defoliate plants or suck them dry of sap; on the other, insects may be vital for pollination. Some bacteria and fungi cause devastating diseases of plants, whereas others are vital to plant nutrition and health. Viruses can debilitate or kill both plants and herbivores. In this chapter, we examine some of these other garden inhabitants, starting with the most generally conspicuous and then working our way to the smallest.

Animals

Animals include mammals, birds, reptiles, amphibians, insects, spiders, worms, slugs, and many other organisms that do not photosynthesize but that have bodies composed of more than one cell. Included in this group are the more familiar animals readily visible to the naked eye, but also the less conspicuous—though equally significant—nematodes, hairworms, rotifers, symphylans, tardigrades, and numerous other minuscule or even microscopic animals little known or noticed by most gardeners.

In a garden ecosystem, animals act as herbivores, organisms that eat plants or parts of plants; as carnivores, most commonly, animals that eat other animals; or as detritivores, organisms that consume decomposed organic material. These functions can be good or bad from a gardener's point of view. For example, herbivory is certainly detrimental when swallowtail butterfly (*Papilio*) caterpillars eat bishop's flowers (*Ammi majus*) in the garden, but herbivory can also have positive consequences, as in the cases of nectar-consuming pollinators and weed-devouring caterpillars. Carnivory is detrimental when crab spiders catch and devour pollinating honey bees, but it is certainly beneficial when carabid beetles consume caterpillars, weevils, or slugs. Detritivory, however, is almost always beneficial: dead animal and plant material is digested and excreted as nutrient-rich feces or incorporated into the bodies of detritivores, such as pill bugs, which themselves can become food sources.

Mammals

Mammals are generally the largest animals found in the garden, although their overall influence is probably minor compared to that of worms, birds, and, especially, invertebrates such as insects, spiders, and their relatives. Locally or under particular circumstances, however, activities of mammals can significantly impact the well-being of garden plants.

Moles (most commonly *Scalopus aquaticus*), shrews (family Soricidae), and hedgehogs (*Erinaceus*) feast largely on insects, many of them harmful to plants. Gardeners, however, generally regard subterranean moles as pests because their burrowing activities uproot plants and leave unsightly ridges and mounds. Moles are especially problematic in North American gardens, and creative although often-hazardous countermeasures have been devised. These include using ultrasonic noise generators; pumping poisonous gas into their burrows; planting toxic castor beans (*Ricinus communis*) in hopes that moles will eat their roots; placing sticks of chewing gum where moles will eat them and, supposedly, choke or develop terminal constipation; placing motion-activated spiked traps above mole tunnels; and even detonating dynamite. However, before resorting to such measures, gardeners may want to weigh the damage moles do against the huge numbers of harmful insects that they eat, especially larvae of garden pests such as Japanese beetles (*Popillia japonica*).

Bats and their relatives are fairly uncommon in most temperate or cool regions, but can be found in significant numbers near caves and in subtropical and tropical regions. Most bats are carnivorous and may consume large numbers of night-flying insects. Others, mostly tropical species, eat large numbers of fruits or the pollen and nectar of night-blooming plants, frequently pollinating plants in the process.

Rabbits and their relatives are major pests of gardens throughout North America, Australia, and Europe. They are especially fond of tender, leafy greens and seedlings, although rabbits will also eat root crops. Steve Salt once lost an entire crop of peanuts to rabbits in a Kentucky garden. Many people have heard of the near-devastation of the Australian countryside that resulted when European rabbits (*Oryctolagus cuniculus*) were released there in the nineteenth century. Although this plague of rabbits has been lessened, it continues to be a concern. How can one control rabbits in the garden, aside from biological warfare, hunting, or a pack of guard dogs? Less dramatic, but often less effective, countermeasures include installing small-mesh hardware-cloth fences 3 feet (90 cm) above-ground *and* 1 foot (30 cm) underground; building electrified mesh fences; wrapping wire mesh around trunks of perennials; and spraying or sprinkling hot pepper extract, coyote urine, bloodmeal, bonemeal, or wood ashes around plants to be protected.

Rodents (mice, squirrels, rats, gophers, and their kin) can cause locally severe damage in gardens (plate 29). Rodents gnaw trunks and roots of woody perennial species and may eat maturing seeds, such as the heads of sunflowers (*Helianthus annuus*). The greatest damage worldwide from rodents, though, is in their voracious destruction of stored grains and other crops. Steve Salt discovered one spring that rodents had gnawed into and eaten the seeds out of more than five hundred dried ornamental gourds stored over the winter in an outbuilding. In some underdeveloped areas, as much as one-third of all stored crops are consumed by rodents or fouled by their feces, greatly aggravating already critical food shortages. The destruction of crops, both harvested and in the field, by rats has been worsened in recent years by widespread killing of their primary natural enemies, snakes. Countermeasures against rodents include trapping and setting out poison baits (used with greatest care so as not to also injure curious children or domestic livestock and pets); wrapping wire mesh around trunks of woody perennials; laying sharp-edged gravel around plant bulbs; and keeping mulch and other shelter for mice at least 3–4 inches (7–10 cm) away from bases of tree trunks. Other measures include wrapping aluminum foil around stems of seed-bearing plant heads; applying hot pepper sprays; storing harvested foodstuffs in metal containers; and encouraging, not destroying, snakes.

Carnivores include dogs, cats, raccoons, and other clawed, fanged, generally meat-eating animals. These animals can be both beneficial and detrimental in the garden. With the exception of raccoons' notorious devastation of sweet corn patches, carnivores generally consume very little plant material. Well-trained watchdogs can effectively drive off many nocturnal garden plunderers such as deer (*Odocoileus*), opossums (*Didelphis virginiana*), raccoons, and rabbits. But dogs are also notorious tramplers and diggers—often in pursuit of moles and gophers—and are especially fond of excavating large depressions in soft soil such as freshly planted seedbeds (plate 30). Domestic cats have long earned their keep by preying on mice and other rodents. However, cats often scratch and defecate in gardens, which is not only aesthetically displeasing but may also spread internal parasites to other animals—including humans. The only practical way to exclude or to keep in most carnivores is a good fence, especially one that is electrified, although few fences can stop extremely nimble felines. Raccoons seem loathe to pass through densely vining pumpkins because the prickles in the vines may hurt their paws; however, few home gardeners have the luxury of enough space to surround their sweet corn with a minimum of 20 feet (6 m) of pumpkin vines. Other creative—although not always feasible or effective—tactics for dealing with raccoons include bagging each ear of corn, spraying ears of corn with hot pepper oil, and leaving a radio tuned to an all-talk station playing all night in the garden.

Primates such as monkeys, baboons (*Papio*), orangutans (*Pongo pygmaeus*), and their relatives are obviously of little significance in North American and European gardens. However, large bands of these animals often pillage fields, orchards, or gardens in tropical Latin America and parts of Africa, India, and the East Indies. For example, monkeys are major pests in West African cacao (*Theobroma cacao*) plantations. Primates tend to be especially fond of fruits, but they are omnivores and may eat almost anything. They are difficult to control because they are intelligent, aggressive, social animals, and large troops or bands may effectively intimidate unarmed human farmers and gardeners.

Elephants can be massively destructive in areas of Africa where they are abundant, tearing up, eating, and trampling gardens and orchards to oblivion in minutes. No fence can stop them, and few gardeners armed with less than a high-powered rifle dare challenge them. Especially during dry seasons, when irrigated garden patches appear as oases of succulent green in a searing brown landscape, pachyderms may be irresistibly drawn to them. This devastation of food gardens—many of which are critical to the welfare of the local people—presents a real dilemma in balancing the conservation of dwindling wild elephant herds against the welfare of frequently malnourished or even starving subsistence farmers.

Ungulates or ruminants such as cattle (*Bos*), sheep (*Ovis aries*), domestic goats (*Capra hircus*), deer, and their relatives can do substantial damage to gardens. Deer, especially, have become major garden pests in recent decades in the eastern and central United States (plate 31). A combination of suburban encroachment into formerly rural areas, restrictions on deer hunting, and abandonment of farmland to wild vegetation has brought rapidly increasing deer populations into close contact with many landscaped yards and gardens. Perhaps the best defensive tactic—short of eradicating all deer in the area—is to surround a garden with a deer-proof fence. However, a truly deer-proof fence needs to be built in a zigzag, at least 8 feet (2.4 m) high, and leaning outward, with a second fence built parallel to it 3–4 feet (1 m) away. Few gardeners, however, are likely to go to such expense and trouble. Other creative—but frequently futile—defenses include hanging human hair or bars of strongly scented soap and spraying urine of humans, dogs, coyotes (*Canis latrans*), or other carnivorous species about a yard or garden. One new and apparently effective technique to deter deer and other nocturnal animals is to place flashing strobe lights around the garden perimeter. Despite all artificial defenses, deer will eat almost any plant if hungry enough, and well-fed, well-watered, succulent garden plants are often preferable to better-defended and less tasty wild relatives. However, favorite plants that are especially likely to attract deer to your garden include tulips (*Tulipa*), hostas (*Hosta*), irises (*Iris*), okra (*Abelmoschus esculentus*) and other mallows, conifers of all sorts, hollies (*Ilex*), roses (*Rosa*) and related plants, sedums (*Sedum*), beets and Swiss chard (*Beta vulgaris*) and related lamb's-quarters (*Chenopodium album*), and azaleas (*Rhododendron*). And just in case sheep get into your garden, Steve Salt has observed that his sheep, at least, are selectively attracted to sour-tasting plants (for example, sorrel, rhubarb [*Rheum rhabarbarum*] purslane, dock [*Rumex*], and *Amaranthus* species) as well as many of the favorites of deer.

Birds

Worldwide, birds account for much more damage to farm and garden plants than do mammals, although many also do invaluable service as predators on animal pests. Most birds are adapted to a highly mobile life of airborne flight, and thus they need to eat high-energy foods that are low in bulk, such as insects. However, many birds do consume large amounts of plant materials, especially the most energy-rich plant parts, the seeds or fruits. A few species, such as hummingbirds, consume flower nectars. Quite unlike mammalian herbivores, virtually no birds subsist on bulky parts such as leaves and roots.

The birds most generally destructive in gardens are perching birds (order Passeriformes), which include European starlings (*Sturnus vulgaris*), Eurasian blackbirds (*Turdus merula*), finches (family Fringillidae), sparrows (*Passer*), and crows (*Corvus*). Perching birds pick out newly planted seeds from the soil and eat maturing seeds of plants both in the field and after harvest. As one ancient planting rhyme says: "Plant one for the cutworm, one for the crow, one for the blackbird, and one to grow." For decades in China, there was a bounty on sparrows due to the severe losses of grain crops they caused. Perching birds also peck at ripening fruits on shrubs and trees; small red fruits such as cherries are especially attractive. Few home gardeners may be concerned about attacks on grain crops other than sunflowers, but devastation of cherries, blueberries (most commonly *Vaccinium corymbosum*), strawberries, and other fruit crops is another story. Also, many troublesome weeds such as poison ivy (*Toxicodendron radicans*), mulberries (*Morus*), chicory (*Cichorium*), ragweed (*Ambrosia artemisiifolia*), black locust (*Robinia pseudoacacia*), and lamb's-quarters are spread when their seeds are consumed and subsequently defecated by birds.

All the other birds of the world combined probably do not equal the destructiveness of the red-beaked quela (*Quelea quelea*), an African passeriform estimated to number as many as 100 billion individuals. Giant

Figure 6. Birds eat a variety of foods, depending to a great extent on the size and shape of their beaks. (A) Red-tailed hawks (*Buteo jamaicensis*) are carnivores with beaks adapted for seizing prey and tearing flesh. (B) Northern cardinals (*Cardinalis cardinalis*) have heavy cone-shaped beaks well suited for breaking open seeds, although cardinals also eat a wide variety of fruits and insects. (C) American robins (*Turdus migratorius*) are omnivores; they use their long, sharp beaks to seize worms and other small invertebrates, but also eat large amounts of fruit. (D) Downy woodpeckers (*Picoides pubescens*) have heavy, armored beaks capable of hammer-drilling into wood and seizing insects inside. (E) Many wrens (family Troglodytidae) have relatively narrow beaks that they use for probing among the leaf litter or in tiny spaces for insects. (F) Hummingbirds (family Trochilidae) have beaks and long tongues specialized for sipping nectar from deep-throated flowers. Birds are not drawn to the same scale.

flocks of quelas have replaced locusts as the number-one agricultural pest in eastern and southern Africa, where farmers and gardeners make great efforts to drive them off or to trap or kill them, usually in vain.

Passeriforms that attack dusk-flying insects such as mosquitoes and moths are generally welcomed into yards and gardens. These insect-hunters include purple martins (*Progne subis*), flycatchers (family Tyrannidae), and swallows (family Hirundinidae). Martin houses once were a traditional fixture of rural homesteads in much of North America and are still constructed by bird-lovers. Other passeriforms get mixed reviews. For example, robins (American: *Turdus migratorius*; European: *Erithacus rubecula*) hunt both harmful insects and beneficial earthworms.

Ducks, geese, and other web-footed waterfowl (order Anseriformes) are relatively insignificant in most modern North American gardens. But old-time weeder geese are finding renewed use in raising difficult-to-weed strawberries by organic methods, as they selectively eat vegetation other than the berry plants—so long as they are kept out of the berry patch when the fruits are ripe! Nowadays, waterfowl tend to be more of a problem for homeowners who do not wish their ponds fouled or lakeside lawns grazed, trampled, and defecated on. However, domesticated ducks (mallard, *Anas platyrhynchos*, and muscovy, *Cairina moschata*), geese (*Anser*), and swans (*Cygnus olor*) are still welcome inhabitants of many Asian and European farms and gardens, where they control many unwelcome insects, snails, and slugs.

Birds of prey such as hawks, eagles, falcons, and their relatives (order Falconiformes) are day-flying birds that may benefit gardens by preying on rodents and rabbits. However, they are also major predators of snakes, which are generally beneficial in gardens. Old World vultures (family Accipitridae), clean up large dead carcasses, accelerating the cycling of nutrients and reducing the spread of pathogens from diseased animals. New World vultures, or buzzards (family Cathartidae), which perform a similar function in the Western Hemisphere, are members of the order Ciconiiformes and are related to storks and herons. Owls (order Strigiformes) take over the role of air-borne predators at night. Most snakes are relatively inactive at night, whereas rodents and rabbits tend to be nocturnal; therefore, owls are more of an unmixed blessing and garden-ers should welcome them. Unfortunately, populations of many owl species, especially barn owls (*Tyto alba*), are in widespread decline in North

America and Europe, most likely due to habitat destruction (such as loss of gabled barns and hollow trees), pesticide use, forest clearing, and the spread of all-night outdoor electric lighting. Where possible, owl boxes should be maintained to encourage these generally beneficial birds. Cat-lovers take note, however: a favorite snack of many owls is milk-fed kittens venturing out at night.

Birds such as domesticated chickens (*Gallus gallus domestica*), guinea fowl (*Numida meleagris*), and turkeys (*Meleagris gallopavo*) are no longer common inhabitants of North American yards, as they were in the nine-teenth century. Nevertheless, fowl may have a significant impact on gar-dens in areas such as Latin America, where they are still commonly raised by subsistence farmers and others. These birds are nonselective and vora-cious omnivores, eating everything from grubs, bugs, and snakes to toma-toes and weed seeds of all sorts. Thus, they can range from a blessing to a curse in the garden. During the off-season, Steve Salt turns his chickens loose in his garden to clean up weed seeds and insect larvae and eggs (plate 32). However, in the summer growing season, if left in the garden, chickens avidly attack almost any round vegetable including beets, onions, cucum-bers, melons, and, especially, tomatoes; chickens also scratch up the soil around carrots and other plants, often uprooting them. Wild turkeys also may damage gardens, especially in wooded areas of eastern North America, although in Steve Carroll's yard they tend to favor the more eas-ily accessible seed spilled from bird feeders. One gardening friend reported, however, that a flock of wild turkeys saved his pumpkin patch by essen-tially exterminating the squash bugs (*Anasa tristis*) threatening his plants.

Frogmouths (family Podargidae), nightjars (*Caprimulgus europaeus*), whip-poor-wills (*Caprimulgus vociferus*), and other night-flying, beakless, insect-eating birds are often irritatingly noisy but nonetheless beneficial garden visitors, snatching and eating both flying and crawling nocturnal insects. Woodpeckers and their relatives (order Piciformes) are a mixed blessing to the gardener. Some members of this order aggressively prey on herbivorous insects, especially those that tunnel through woody plants. However, the holes that they bore in pursuit of their prey may structurally weaken plants and also provide entryways for pathogenic fungi and bacte-ria. Other piciforms, such as toucans (*Rhamphastos* or *Ramphastos*), are fruit-eaters and may pillage tropical orchards. Pigeons, doves, and their rel-atives (order Columbiformes) are of little consequence in most North

American gardens, but they can be significant pests in some Mediterranean areas, where they consume fruits, grains, and other crops. Hummingbirds (family Trochilidae) are nectar-feeding pollinators of a great variety of ornamental and native plants such as trumpet creepers (*Campsis radicans*), coral bells (*Heuchera sanguinea*), or species of *Lobelia* or *Monarda* and many orchids. These birds do little or no harm, however, and are welcomed to most gardens. Their hovering or flittering flight is fascinating to watch and adds to the pleasure of growing the flowers they favor.

Reptiles

Many gardeners shudder at the thought of reptiles in their gardens, but these animals are generally beneficial. Snakes are often more efficient predators of rodents than are cats or dogs, being silent, generally odorless (unless frightened or seeking a mate), and able to follow rats and mice into their holes and tunnels. To encourage snakes to take up residence, provide appropriate habitat such as loose rock piles, wood piles, or old sheds with gaps or holes in the walls—and, of course, resist the urge to kill them on sight. Good snake habitat is also good rodent habitat, though, so one is faced with a bit of a dilemma.

In North America, a variety of nonpoisonous snakes such as black or eastern racers (*Coluber constrictor*), bullsnakes (*Pituophis catenifer sayi*), brown snakes (*Storeria dekayi*), and garter snakes (*Thamnophis*) prey on insects, slugs, snails, frogs, and, especially, rodents (plate 33). Poisonous snakes such as rattlesnakes (*Crotalus*) and copperheads (*Agkistrodon contortrix*) are very efficient predators of pests, but are definitely unwelcome in most gardens. Brightly colored nonpoisonous king snakes (*Lampropeltis*) are doubly beneficial; not only are they voracious predators of rodents, but they also prey on rattlesnakes.

Lizards, turtles, tortoises, and other reptiles are relatively infrequent visitors to most temperate-region gardens. Some Floridian and Gulf Coast gardeners, however, have been startled to encounter American alligators (*Alligator mississippiensis*) lumbering through their yards, and Steve Carroll once had to extricate a very large snapping turtle (*Chelydra serpentina*) that was determined to lay her eggs in the soft soil of his Massachusetts flowerbed (plate 34). In subtropical and tropical regions, lizards can be significant predators of small rodents and insects. In desert regions of the American Southwest, horned toads (actually lizards,

Phrynosoma) and other lizards are frequent visitors to yards and gardens, attracting in turn their deadly avian nemesis, the greater roadrunner (*Geococcyx californianus*). In India and the West Indies, geckos (*Gekko*) are traditionally welcome in homes for controlling mosquitoes.

Amphibians

Salamanders and frogs are most commonly encountered in gardens only in damp areas or close to ponds or other calm waters, although tree frogs are often found in wooded areas (plate 35). Toads (*Bufo* and relatives) are widespread, however, as their dry, thick skin and short, stout legs with unwebbed feet make them adapted for life in drier environments than can be tolerated by frogs. Primarily nocturnal, toads are usually welcomed by gardeners wherever found, as they are strictly carnivorous (plate 36). One active toad can consume hundreds of slugs, flies, grubs, cutworms, and other pests in a single night. Many gardeners construct toad houses or buy commercially made models in hopes of enticing these warty insect-destroyers to take up residence. Like frogs, however, toads require free-standing water for laying their eggs. Shallow ponds or frequently replenished pans of water with their brims at ground level will make a garden a friendlier place for toads. Incidentally, toads are surprisingly long-lived, reaching four to fifteen years in age under favorable conditions, and may benefit a garden for a long time once they take up residence.

Fish

Water gardening is the practice of creating and maintaining artificial ponds, streams, or wetlands or managing natural ones, together with their associated plants and organisms, in much the same fashion as terrestrial gardens. To maintain healthy garden ecosystems, those who maintain aquatic or wetland gardens need to pay attention to the numbers and species of fish in their ponds. Fish eat algae; they prey on frogs, insects, and other pond or marsh animals; or they browse on aquatic plants. Stocking and maintaining desired species of fish in water gardens—or helping fish maintain desired plant and insect species and control unwanted ones—requires attention to water chemistry and temperature, depth and flow rates, shade versus sun exposure, shelter for organisms, behavior patterns of fish, and a host of other factors. Does it sound difficult? It's not really, and water gardening is a fascinating and increasingly popular form of landscaping.

In more traditional gardening, fish are most often encountered in the form of fish emulsion used as a fertilizer and plant tonic by organic gardeners. However, live fish may impact terrestrial garden ecosystems by preying on eggs and tadpoles of frogs and toads and on larvae of insects such as mosquitoes, the adults of which may otherwise travel to nearby gardens. Also, fish may play significant roles in the ecology of traditional rice paddies, besides serving as food for paddy farmers.

Insects

The ecological roles of insects in the garden are powerful and complex, and the influence of insects is generally rivaled only by that of the microscopic nematodes, bacteria, and fungi. Insects exist in such great numbers, grow and reproduce so quickly, and have such voracious appetites that their consumption of garden plants usually far exceeds that by all other herbivores put together. In addition, many serious bacterial, viral, and fungal diseases are carried from plant to plant on insect mouthparts or in insect guts. In fact, most viral diseases of plants can spread solely by means of insects. And, of course, insects that bite, sting, or burrow into humans' skin are among the greatest literal pains endured by gardeners. Predatory insects, however, are one of the most effective means of controlling harmful insects, mites, and other pests. Many plants depend on insects for pollination, and herbivorous insects are the primary controls of some noxious weeds. To complicate matters further, from a gardener's point of view an insect can be harmful at one life stage or under one set of circumstances but beneficial at another.

Insects are arthropods (phylum Arthropoda), a group that also includes arachnids (spiders and mites), symphylans, centipedes, millipedes, and crustaceans (pill bugs, crabs, and lobsters). Arthropods have in common a tough but flexible outer shell (an exoskeleton) in their adult forms, but they lack an internal framework or skeleton like mammals. Exoskeletons enable arthropods to survive powerful blows, extreme temperatures, and dehydrating winds and to resist penetration by many chemical substances. However, if an exoskeleton is breached, as by a sharp object, arthropods may rapidly die from dehydration, leakage of critical body fluids, or invasion by pathogenic microbes. Many immature arthropods lack the hardened exoskeleton and are therefore quite vulnerable to injury, predators, and disease. In apparent compensation for this high infant mortality rate, most arthropods produce prodigious numbers of offspring that grow very rapidly (consuming huge amounts of food as they do), thus accounting for much of their damage to garden plants.

Insects are distinguished from other arthropods by having three distinctive body segments: a head with mouthparts; a muscular midsection, the thorax, to which are attached six legs and the wings or shell (if present); and a posterior abdomen containing digestive, reproductive, respiratory, and other organs. Insects have no lungs and lack a true circulatory system, having only a primitive heart that acts somewhat like a sump pump, an open-ended pump that sucks liquid from an open reservoir and dumps it elsewhere. Oxygen enters insects and carbon dioxide waste exits through numerous tiny pores (spiracles) in the body surface. Although insect brains are very small and rudimentary, their decentralized nervous systems respond extremely quickly to external stimuli, giving them lightning-quick reflexes as well as high sensitivity to many neurotoxins. This sensitivity is taken advantage of by farmers and gardeners in the application of neurotoxic pesticides such as the organophosphates, which were originally developed during World War II as chemical warfare agents.

Insects are cold-blooded (more properly, ectothermic), meaning that their body temperatures are not effectively internally regulated but tend to fluctuate with the external environment. (Some bees and other social insects, however, can raise their body temperature and collectively regulate the temperature inside enclosed colonies by heat-generating or cooling activities.) One consequence of this limitation is that insects tend to be physically and metabolically much more active at higher temperatures and relatively sluggish at lower ones. Because they do not expend energy to regulate their body temperature, however, insects can survive on a small fraction of the calories that would be required by warm-blooded (endothermic) animals of similar body mass. Furthermore, during periods of prolonged cold or severe drought, many insects can enter a deep dormancy with very low requirements for oxygen and nutrients. In many cases, insects can satisfy their need for water largely by metabolizing carbohydrates to carbon dioxide plus water, which they conserve and use internally. This ability enables their survival during droughts, in deserts, or in very dry environments such as grain-storage silos or garden seed packets.

Insects also tend to reproduce and develop at a pace proportional to their environmental temperature. Thus, at one temperature a given species may take several weeks to pass through one life cycle, but at a temperature 15–20°F (9–12°C) higher may produce two to three generations in the same time span. This is one reason why insects in continuously warm regions, such as the tropics, tend to be more severe pests than those in temperate or polar regions. Other reasons are that winter cold seasons tend to kill insects or drive them into dormancy as well as deprive them of actively growing plant food sources.

Another biological feature important to the ecology of garden insects is their multiphase life cycle. Some insects go through complete metamorphosis (from the Greek for "changing form"). The juveniles of these insects look very different from the adults. They are usually soft-bodied organisms such as caterpillars or grubs that grow by stretching or molting their soft skins, while voraciously consuming plant or animal tissues. Insects of this type have to transform into adults through the dormant stage of a pupa within a cocoon. The adult insect may look little or nothing like its juvenile form and generally has very different dietary habits. Adult cicadas (family Cicadidae) and mayflies (order Ephemoptera), for example, eat little or nothing and are not much more than short-lived reproductive machines, whereas juveniles of the same species are long-lived voracious grubs. Many butterflies and moths sip nectar from flowers while serving as pollinators, whereas their juvenile forms (caterpillars) are leaf-eating herbivores. Adult beetles are often carnivorous predators, whereas many juveniles are dung- or root-eating grubs or parasites of animals. Many adult flies are liquid-sippers, whereas their juvenile forms (maggots) ingest decaying plant or animal materials.

Other insects go through incomplete metamorphosis, in which juveniles look more or less like miniature versions of the adults, albeit usually with softer exoskeletons and somewhat different body proportions. These insects grow through a series of molts, essentially zipping open their exoskeletons as they become too restrictive for the growing juveniles, emerging as somewhat larger adult look-alikes. The number of times an insect molts defines the number of life phases, or instars, that the insect passes through, with the final instar being the reproductively capable adult. In insects that undergo incomplete metamorphosis, such as grasshoppers and locusts (order Orthoptera) and true bugs (order Hemiptera), the adult and juvenile forms generally have similar dietary habits and behavior—aside from the mating habits of the adults, of course.

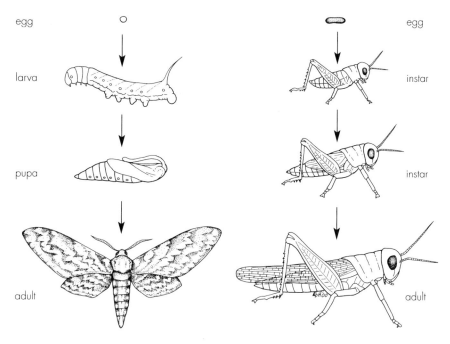

egg

larva

pupa

adult

egg

instar

instar

adult

Figure 7. Insects can be described based on how they change from egg to the adult stage. In complete metamorphosis, characteristic of beetles, butterflies, and moths (left), the egg, larva, pupa, and adult are substantially different from each other. In incomplete metamorphosis, characteristic of true bugs and grasshoppers (right), reproductively immature stages (instars) resemble adults.

A few insects, notably aphids, are capable of parthenogenesis (reproduction without sex, or cloning). Females in these groups are able to clone themselves without fertilization by a male, bearing young that are genetically identical to themselves via a live birthing process. Parthenogenesis can result in sudden and spectacular population explosions under favorable conditions, as all too many greenhouse gardeners and houseplant growers can testify.

Insects usually have acute senses of smell and are able to locate suitable food and mates from great distances by tracking extremely faint odors. This is how a squash bug can target your lone pumpkin plant from miles away, and why hornets and bees may be attracted to you if you wear a floral perfume. Beekeepers will testify that honey bees can sense mammalian body odors that signal emotional status, and bees are much more prone to attack a person who is angry or fearful. Insects generally communicate with one another by means of pheromones (signaling scents). For example, a male gypsy moth (*Lymantria dispar*) can locate a single sexually receptive female by her scent from as far away as 2 miles (3.2 km) downwind.

Many insects have sharp eyesight as well, although their views are split into numerous images by compound lenses. Compound eyes are especially sensitive to motion, as anyone who has tried sneaking up on a butterfly knows. Many insects also have excellent color vision and are selectively attracted by particular colors; some can even detect ultraviolet wavelengths invisible to humans. Apple leafminers (*Phyllonorycter crataegella*), maggot flies (*Rhagoletis*), and codling moths (*Cydia pomonella*) are attracted to red; tarnished plant bugs (*Lygus lineolaris*) and many night-flying moths to white; mosquitoes to ultraviolet; cucumber beetles (spotted: *Diabrotica undecimpunctata*; striped: *Acalymma vittatum*), flea beetles (subfamily Halticinae), and greenhouse whiteflies (family Aleyrodidae) to yellow; and western flower thrips (*Frankliniella occidentalis*) and many flies to blue. This attraction of insects to particular hues may relate in some cases to the colors of their preferred natural food sources—yellow cucurbit blossoms in the case of cucumber beetles, red fruits in the case of codling moths, white night-blooming flowers for many moths—but is not understood in other cases.

Of great relevance to gardeners is the fact that insects often have specific dietary habits, both as to what they eat and how they go about eating it. Virtually all insects have preferred foods—those that they will feed on first if available—but some insects will also feed on a wide range of other plants. Others are particularly selective, consuming only one or a few closely related plant species and may starve if deprived of their preferred food source. And for any given insect species, many plants will not sustain the insects or are actually harmful; these are avoided except under starvation conditions (and sometimes even then). All too often, garden plants, especially those of edible crops, fall into the preferred-food category. Nutrient and moisture levels of domesticated plants are often higher than in wild or untended plants, and toxins, thorns, hairs, tough shells, and other plant armament have been reduced by breeding to make them more fit for human use. For example, the Colorado potato beetle (*Leptinotarsa decimlineata*; plate 37), a serious pest of the crop worldwide, was a minor herbivore of wild *Solanum* species in western North America until commercial potato growing began in the area. The beetle discovered potatoes

(*Solanum tuberosum*) to be a preferred food and soon spread to other potato-growing areas.

Eating styles among insects may be broadly classified into chewing or sucking. Chewing insects tear off, chew, and ingest chunks of tissues using pincerlike mouthparts. This group of insects includes beetles, grasshoppers, cockroaches, earwigs, thrips, and the juveniles of moths and butterflies (caterpillars). Sucking insects have syringelike or lapping mouthparts and are restricted to a diet of liquids or very soft materials. A diet of goo may sound restrictive, but it may encompass many readily available nutrient-rich foodstuffs such as plant cell saps, animal blood, floral nectars, and juices oozing from decaying or injured tissues. Sucking insects include adult flies, mosquitoes, butterflies, and moths, as well as aphids and true bugs at all life stages. How insects eat is of relevance for their control; for example, sucking insects must be controlled with contact or systemic insecticides because they are not affected by surface-coating substances that depend on ingestion for their toxicity.

A vast number and tremendous diversity of insect species exist worldwide. Indeed, there are more known species of insects than all other species of macroscopic organisms combined. The group having the greatest species richness is the beetles (order Coleoptera), which includes an estimated 28 percent of all animal species. Beetles pass through complete metamorphosis. The larvae (grubs) are generally subterranean and feed on plant roots and tubers, although some prey on other soil organisms (plate 38). After pupation, the adult beetles emerge into the upper world, where most prey on other insects, slugs, worms, and other invertebrates; some, however, are serious herbivorous chewing pests. Ravenous Japanese beetles and chafers (or scarab beetles, family Scarabaeidae) do most of their damage as grubs in the soil. Wireworms (family Elateridae; known in their adult form as click beetles) prey on immature seeds. Cucumber beetles not only chew on leaves, roots, and flowers of cucurbits and other plants, but also transmit deadly bacterial blight to cucumbers and muskmelons. Bark beetles (family Scolytidae) burrow under the bark of many perennial woody plants and transmit diseases such as Dutch elm disease. The voracious Mexican bean beetles (*Bruchus*), which resemble pink ladybird beetles, feed on leaves and flowers of a great many hosts besides bean plants. Tiny, black, nimble flea beetles riddle leaves of a wide range of plants, although their preferred hosts seem to be eggplants (aubergines) and related *Solanum* species. Beneficial beetles include coccelinids such as the well-known ladybird beetles (or ladybugs, *Hippodamia convergens* and *Cryptolaemus*); spider-mite destroyers in the genus *Stethorus*; rove beetles (family Staphilinidae) that prey on aphids; and the aggressive, pincered ground or carabid beetles (family Carabidae) that attack caterpillars, weevils, slugs, grasshoppers, and even, on occasion, small birds or mammals.

Caterpillars are the larval stage of moths and butterflies (order Lepidoptera). They are probably the most generally plant-destructive group of herbivorous insects, although their consumption of weeds may indirectly benefit garden plants (plate 39). Prominent North American lepidopteran pests include the small white cabbage butterfly (or, mistakenly, cabbage moth; *Pieris rapae*), which in its caterpillar stage is the single most destructive pest of crucifers such as cabbage, broccoli, cauliflower, and rape (*Brassica napus*). Peach stem borers (*Anarsia lineatella*) are major pests of stone fruit trees such as apricots, peaches, and cherries. Codling moths are some of the most important causes of wormy apples. Sod webworms (*Crambus*) damage turfgrass, and European corn borers (*Ostrinia nubilalis*) are a major cause of wormy ears of corn/maize. Gypsy moth larvae and tent caterpillars (*Malacosoma*) cause extensive damage to a wide range of woody perennial plants. The spectacularly large and ugly larvae (hornworms) of the hawkmoths (*Manduca*) attack tomatoes and other solanaceous plants as well. A single hornworm can defoliate a large tomato plant. Cutworms (family Noctuidae) are probably rivaled only by damping-off fungi as a cause of death of young seedlings. Directly beneficial caterpillars are less well known, but include monarch butterfly (*Danaus plexippus*) larvae, which feed on milkweed, and others that selectively attack weeds.

Of course, regardless of the damage done by caterpillars, many people enjoy the colorful dancing confetti of butterflies in flight, and many gardeners grow plants such as butterfly bush (*Buddleia davidii*), butterfly weed (or milkweed; *Asclepias*), and purple coneflower (*Echinacea purpurea*) to attract butterflies to their gardens (plate 40). Moths are generally nocturnal, and many night-blooming, aromatic, nectar-rich plants such as thorn apples (*Datura*) are moth plants. A gardener may thus deliberately introduce plants to sustain caterpillars that will develop into desired butterflies, but will want to take care not to attract butterflies or moths whose not-so-attractive larval stages may devastate desirable garden plants (plate 41).

Grasshoppers, locusts, crickets, and their relatives (order Orthoptera) are generally not as great a problem in home gardens as are moth and butterfly larvae and some other insect groups, but under particular conditions—such as during warm, dry weather following a mild winter—they can become devastating. Locust swarms of the dry tropics and subtropics and cricket and grasshopper swarms in the drier, warmer parts of the United States and other warm-temperate regions are infamous for their mass destruction of plant life over large regions. Orthopterans are voracious chewing herbivores whose large size and strong mouthparts enable them to consume relatively tough plant parts generally left alone by other insects—including older stems, hairy leaves, and hard-shelled fruits. Under normal conditions, insects in this group mostly attack their preferred hosts, members of the mint and grass families, causing partial defoliation. However, when populations mount to high levels, orthopterans will attack almost any plant, resulting in serious damage. Swarming occurs when millions or even billions of individuals congregate and migrate en masse, devastating virtually all plant life in their path. Fortunately, the awesome swarms of the late nineteenth and early twentieth centuries have not recurred. This reprieve is possibly due to a combination of factors including human disturbance of orthopteran breeding grounds and control of nascent swarms at earlier and more manageable stages by spraying of insecticides or spreading of baits laced with poisons or parasites.

Many people refer to all insects as "bugs," but properly this label belongs only to insects of the order Hemiptera, the true bugs and close relatives such as leafhoppers, a group of approximately 40,000 species. True bugs are relatively soft-bodied insects that have piercing-sucking mouthparts, especially in juvenile life stages, and that undergo incomplete metamorphosis. Most are herbivores, many are serious plant pests, and some are aggressive predators of other arthropods. Like sucking insects in general, some bugs are vectors of viral and other plant diseases, thereby greatly aggravating the direct feeding injury they do to garden plants. Notorious North American hemipteran pests include leafhoppers (several genera of the family Cicadellidae or Jassidae), which do most of their damage by transmitting viral diseases from plant to plant. Tarnished plant bugs (*Lygus*) especially attack tree fruits but have a wide range of alternate hosts. Chinch bugs (*Blissus leucopterus*) attack grasses and grain crops. Squash bugs, probably the single most devastating pest of pumpkins and squashes, will feed on other cucurbits and even other types of plants if preferred hosts are absent. And, then, there are the infamous bed bugs (*Cimex lectularis*), which torment many gardeners at night. Beneficial bugs include predators of insects, such as assassin bugs (various genera of the family Reduviidae), toad bugs (*Gelastocoris*), and spined soldier bugs (*Podisus*). Ambush bugs (*Phymata*) are also aggressive predators, but they kill so many beneficial insects, such as honey bees, that they are generally regarded as pests in the garden.

Gardeners most often encounter insects of the order Homoptera (aphids, whiteflies, scales, cicadas, and relatives) on houseplants or in greenhouses. However, under certain conditions, such as mild temperatures combined with high humidity and calm air, they can be major plant pests outdoors as well. Homopterans have sucking mouthparts and the softest bodies of any insects, and thus are very vulnerable to predators, contact pesticides, and injury. However, they are capable of prodigious rates of growth and reproduction under favorable conditions, so even a few survivors can multiply into massive infestations seemingly overnight. Some homopterans, such as aphids (plate 42), are capable of parthenogenesis. Homopterans prominent in North America include many species of aphids (family Aphididae); greenhouse whiteflies (*Trialeurodes vaporariorum*), probably the single most severe pest of greenhouse and indoor plants; and scale insects of many genera, which are major pests of a wide range of perennial woody plants. Homopterans also include periodical cicadas (*Magicicada*). As the noisy, short-lived adults that emerge aboveground once every thirteen (in southern regions) or seventeen years (in northern regions), cicadas cause relatively little damage; however, as juveniles they feed for many years on roots, causing often unrecognized debilitation of perennial plants. Homopterans rarely kill plants outright or cause the conspicuous damage characteristic of chewing insects, but they stunt plants by draining nutrients and often transmit viral diseases. The green peach aphid (*Myzus persicae*) is reputed to transmit more than a hundred plant viral diseases. Honeydew excreted by many aphids and scale insects also promotes colonization of plant leaves by bacteria and fungi, some of which are relatively harmless, such as sooty molds, but many of which are pathogenic (plate 43).

Thrips (order Thysanoptera) are tiny, leaf-mining insects with broad diets, which are quite notorious as vectors for plant viral diseases. Their

Other Garden Inhabitants

attacks may also severely injure or even kill small seedlings by destroying their leaves. Thrips also destroy petals of flowers. Especially susceptible are onions and other alliums, grasses, lilies, legumes, and flowering ornamentals such as chrysanthemums.

Insects of the order Diptera include many unfortunately prominent pests of animals such as mosquitoes (family Culicidae), houseflies (*Musca*), and diverse biting flies (which do not actually bite, but rather stab—quite painfully). In the plant world, dipterans tend to be more beneficial because the group includes important pollinators, as well as predators and parasites of other insects and harmful organisms. The order Diptera also includes some significant herbivorous pests, however, especially in the larval stage. Adult dipterans generally are liquid-feeders; larvae may feed on liquids or chew solids, but tend to prefer soft materials such as decaying tissues.

Dipterans that are significant pests of North American plants include mosquito-like craneflies (*Tipula*), whose subterranean larvae feed on grasses and on alternate hosts planted into freshly tilled sod. Midges and gall midges (families Ceratopogonidae and Chironomidae) are tiny insects that burrow into and lay eggs in stems, fruits, and seeds and whose adults bedevil gardeners as pesky no-see-ums. Hessian flies (*Mayetiola destructor*) are the foremost insect pest of wheat. Onion bulb flies (*Eumerus strigatus*) and onion maggots (*Delia antiqua*) are major pests of bulbs such as tulips (*Tulipa*), narcissus (or daffodils), amaryllis, onions, and garlic (*Allium sativum*). The infamous Mediterranean fruitfly (or medfly, *Ceratitis capitata*) devastates a large array of tropical and subtropical tree fruits and has occasioned numerous controversial aerial pesticide spraying campaigns in California and Florida (plate 44). Other dipteran pests include various fruitflies (family Tephritidae, also called Trypetidae), which mainly attack harvested fruits in storage, and leaf miners (family Agromyzidae), which burrow into the leaves of many plants.

Beneficial dipterans include many major parasites and predators of caterpillars and other harmful insects. Parasitic flies include tachinid flies (family Tachinidae), phorid flies (family Phoridae), and bombylinid flies (family Bombylinidae). Beneficial predatory flies include robber flies (family Asilidae) and syrphid flies (or hoverflies; family Syrphidae). Also, dipterans including hoverflies, blow flies (*Phormia*), bottle flies (family Calliphoridae), and houseflies (*Musca domestica*) are important pollinators of plants with flowers that lack nectar or that are too small for bees

67

and wasps. Unfortunately, the larvae of some dipteran pollinators are destructive herbivores. Dipterans definitely receive mixed reviews from gardeners.

From a gardener's perspective, the order Hymenoptera (ants, bees, hornets, wasps, and their relatives) is the most generally beneficial of all insect groups, and control measures for pests should be conducted so as to minimize injury to these insect friends. Hymenopterans pass through complete metamorphosis; larvae are generally carnivorous or parasitic on other insects, whereas adults have the most versatile mouthparts of any insects and can chew, lap, and suck. Preferred food sources for most adult hymenopterans are floral nectars, although pollens, sweet fruits (especially of the rose family), and tender leaves are also eaten. Their beneficial activities far outweigh any harm done by their modest herbivory—although a person stung while picking a pear or raspberry in which yellow jackets (*Vespula*) are feeding may not agree! Some ants are aggressive carnivores and can attack and dismember insects, reptiles, birds, and even mammals with their powerful mandibles.

Humans and many other animals fear hymenopterans for their ability to administer painful stings, but most species rarely attack except to defend their homes or if harassed. Stinging is also used to paralyze other insects, usually soft-bodied larvae such as caterpillars, which the hymenopteran adults transport back to their nests to feed to their own larvae. Parasitoid wasps such as ichneumonids (family Ichneumonidae), trichogrammatids (family Trichogrammatidae), and braconids (family Braconidae) are among the most significant controls on populations of moths and butterflies, flies, and true bugs. They lay their eggs on larvae of these insects, sometimes after immobilizing them by stinging. When the parasitoids' eggs hatch, their larvae consume the flesh of the host larvae. Honey bees (*Apis mellifera*), bumble bees (*Bombus*), mason bees (*Osmia*), and leaf-cutting bees (*Megachile*) are probably the most widespread and significant pollinating insects, and successful seed and fruit production by a wide array of plants depends on them (plate 45). Some tiny wasps are critical pollinators of plants with flowers so small as to give access to virtually no other insects.

Members of the order Odonata (dragonflies, damselflies, and relatives) are generally beneficial. As larvae they prey on aquatic insects and other creatures, and as swift adults they attack flying insects on the wing.

Dragonflies are especially useful for control of both larval and adult mosquitoes and are thus of particular benefit to water gardeners or persons living near ponds or wetlands. Interestingly, the incredibly agile and voracious dragonflies are considered by evolutionary biologists to be some of the most primitive insects, virtually unchanged from very old fossil forms.

Lacewings (families Hemerobiidae and Chrysopidae), alderflies (family Sialidae), antlions (family Myrmeleontidae), dobsonflies (family Corydalidae), and relatives (order Neuroptera) are generally predators of other insects. In particular, lacewings and their larvae, sometimes called aphid lions, prey voraciously on soft-bodied insects. Lacewings are commercially available from insectaries and garden supply houses for use in biological control of many garden pests (see chapter 6).

Springtails (order Collembola) are tiny insects that generally live in moist habitats, where they feed on decaying vegetation and soil fungi. Their common name refers to the distinctive lever or flipper under tension on the abdomen that, when released, abruptly flips the insect a great distance as a means of escaping predators. Springtails occasionally become locally important pests of leafy legumes, lettuce, or tender seedlings. Control is as for other soft-bodied insects, that is, use of insecticidal soaps, oil sprays, contact insecticides, and good garden sanitation.

Earwigs (order Dermaptera, also called Dermoptera) are a relatively small group of mostly nocturnal chewing insects. Generally, they are not significant pests but may prey on tender fruits and flower petals and occasionally seriously damage cutting flowers. Good garden or greenhouse sanitation is usually adequate for control, as earwigs hide during the daytime under rocks, boards, or other objects. Another suggested control measure is laying down wet newspapers in or near the garden at night, then collecting the earwigs hiding underneath in the morning.

The infamous cockroaches (order Dictyoptera, also known as Blattaria) are primarily pests of human habitations, doing little damage in the garden, even though they are voracious and highly adaptable omnivores. Gardeners may become alarmed at the sight of related wood roaches (*Parcoblatta*) in the garden, but these are harmless insects and should be ignored. Praying (or preying) mantises (family Mantidae) are usually classified as dictyopterans, although sometimes as orthopterans with the grasshoppers and crickets—or even given their own order (Mantodea). Mantises were formerly popular for biological control of insect pests in gardens, but they have fallen out of favor because they prey indiscriminately on both harmful and beneficial inhabitants of the garden.

Arachnids

Spiders (order Araneae), daddy-longlegs (family Phalangiidae), mites (order Acarina), and their relatives have only two body sections (an enlarged abdomen and a relatively small head-thorax) rather than the three of insects. They also use fanglike cheliceras rather than mandibles for grabbing their prey. Arachnids cannot actually chew. They generally first immobilize prey by injecting a poison, then drool or inject digestive enzymes onto or into their catch, and suck or lap up the digested materials. Alternatively, arachnids simply pierce their victim (either plant or animal) and suck out its fluid contents.

Spiders are generally of benefit to humans in the garden, and usually in the house, for that matter—cobwebs notwithstanding (plate 46). Spiders energetically pursue, seize, and consume large numbers of crawling insects (including the otherwise seemingly indestructible cockroaches), or trap flying insects with their intricate, often sticky webs. Thus, webs of the large, fast-moving, bright yellow-and-black garden spider (*Argiope aurantia*) should be a welcome sight for gardeners. At least in North America, many garden spiders and daddy-longlegs are actually quite poisonous. They pose little threat to humans, however, because they are unable to penetrate human skin with their very short fangs and, in any case, they virtually never attack mammals. North American poisonous spiders that are significantly hazardous to humans are not commonly found in gardens; these include the large black widow (*Lactrodectus mactans*), with her distinctive reddish orange hour-glass marking; the plain-looking, rather hairy brown recluse spider (*Loxosceles reclusa*); and the large, fast-moving hobo spider (or aggressive house spider, *Tegenaria agrestis*). There are numerous garden and house spiders in Great Britain, but none are known to be poisonous to humans—although there are rumors of recent venomous immigrants.

Most familiar to the majority of people are the orb web spiders (family Argiopidae), of which the common garden spider is one example. These spiders weave large, vertical, conspicuous, open webs with discrete strands of fibers. Probably more significant in control of insect populations, however, are the money spiders (family Linyphiidae), which are usually small and inconspicuous and weave dense, often gossamer-like webs of many shapes in nooks and crannies of rock formations, piles of debris, rumpled or folded leaves, branching plants, and houses. Cobwebs are most often the webs of money spiders. Spiders that do not build webs include the wolf spiders (family Lycosidae), large, hairy, and swift nocturnal hunters that chase down their prey and paralyze them by injection of venom (some tropical species are dangerous to humans); the crab spiders (family Thomisidiae), which ambush pollinating insects inside flower heads, usually in the daytime; and the jumping spiders (family Salticidae), which, as their name suggests, leap onto prey, usually during daylight.

Mites, ticks (suborder Metastigmata), chiggers (*Trombicula*), and relatives—small, spiderlike arachnids—can range from very beneficial to very destructive. Ticks and chiggers rarely harm garden plants, but feed on the blood of warm-blooded animals such as birds, deer, dogs, and gardeners. Some ticks (families Argasidae and Ixodidae) transmit serious ailments such as Lyme disease, typhus, Rocky Mountain spotted fever, and psittacosis. Ticks and chiggers are especially a problem in wooded or brushy rural and suburban areas with abundant wildlife, attaching themselves to gardeners who brush against plants. Gall, rust, and blister mites (family Eriophyidae) feed on a wide range of plants, causing various deformations of leaves as well as formation of galls and blisters. Spider mites and their relatives (family Tetranychidae) are severe pests of many houseplants and of garden plants in warm, dry conditions. Spider mites pierce leaves and tender young stems and suck out their sap, causing wilting. They also cause a characteristic silvering or bronzing of foliage and tender shoots, defoliation, and often death of plants before the gardener is fully aware that there is an infestation. In heavy infestations, spider mites may construct small gossamer-like webs in nooks and crannies on undersides of leaves, hence their common name. Virtually all plants may be attacked by mites, but especially susceptible are woody perennial plants and members of the celery (Apiaceae or Umbelliferae) and mint (Lamiaceae or Labiatae) families. Ironically, predatory mites (most commonly, family Phytoseiidae) are the major natural controls of plant-parasitic mites. They are also major predators of many insects, both beneficial and harmful (plate 47). Actually, indiscriminate use of miticides may do more harm than good, because predatory mites are often more sensitive to toxic chemicals than are herbivorous mites.

Symphylans

Symphylans (or symphylids) are a group of arthropods whose substantial effect on garden plants has been recognized only recently. Most commonly represented by the so-called garden centipede (*Scutigerella immaculata*), symphylans are actually neither true centipedes nor insects, but rather a class of organisms with some characteristics of both. Indeed, the class name Symphyla means "together with or in between other phyla." The bodies of symphylans resemble stubby centipedes with only twelve sets of legs, and the heads and mouthparts resemble those of chewing insects. Symphylans are most abundant in soils in areas that have been converted recently from woodlands (for example, new subdivisions in wooded areas) or in moist soils rich in incompletely decomposed organic matter. Root-feeding symphylans stunt plants by aggravating nutritional deficiencies, increasing susceptibility to drought stress, and facilitating entry into plants of pathogenic soil-dwelling fungi through the wounds that they cause.

Centipedes and Millipedes

Both centipedes and millipedes are found worldwide, except for polar regions, but are most abundant in humid tropical, subtropical, or warm temperate regions. Centipedes (class Chilopoda) don't actually have a hundred legs, but may have nine to forty armored body segments, each with a single pair of relatively long legs. Fast-moving, aggressive, mostly nocturnal predators with poisonous pincers, the more than 4000 known species of centipedes are mostly beneficial in the garden as they hunt and kill insects, slugs, nematodes, earthworms, protozoans, and other garden animals—even, on occasion, small mice and voles. Unlike most arthropods, centipedes care for their young, tending and guarding both their eggs and hatchlings.

Slow-moving and reclusive, millipedes (class Diplopoda) have two pairs of legs on each armored body segment, for a total of as many as two hundred to three hundred legs. The number of legs increases throughout a millipede's lifetime as new segments are added with each molt. The estimated 8000 species of millipedes are mostly detritivores, feeding on decaying plant material, although some are herbivores that prey on living plants. Millipedes ooze a noxious odor when disturbed, but they are generally quite harmless to other animals.

Crustaceans and Mollusks

Crustaceans are represented in gardens by the familiar sow bugs and pill bugs (order Isopoda; sometimes called roly-polies because of their habit of rolling themselves into tight, armored spheres when disturbed). These creatures dwell mostly in damp, shaded locations such as deep leaf litter or underneath loose rocks or boards. Isopods live primarily on decaying plant material. Because they are not harmful to plants, sow bugs and pill bugs might best be simply left in peace. In gardens close to streams or ponds, other crustaceans such as crayfish (order Astacidae; "crawdads" to many) may be encountered, as may their deep, narrow burrows topped by mud structures that resemble chimneys. Unless you are so foolish as to stick a finger into their pincers, these crustaceans are generally harmless.

Some mollusks such as snails and their shell-less cousins, the slugs, can be severe garden pests, especially in the arid western parts of North America (plate 48). The predominance of snails and slugs in arid regions is somewhat ironic, as these creatures prefer to live in damp, cool to mild locations. They hide during the heat of the day under rocks or in other sheltered locations, coming out at night to chew large, ragged holes in leaves and fruits of garden plants or to eat young transplants or newly emerged seedlings right down to the ground. Not only do snails and slugs directly consume considerable amounts of plant tissues, but the wounds that they inflict commonly become entryways for rotting or pathogenic fungi or bacteria that may then spread through the plants, causing even more extensive damage.

Segmented Worms

Segmented worms include earthworms such as the familiar red wrigglers (*Eisenia foetida* and *Lumbricus rubellus*), nightcrawlers (*Lumbricus terrestris*), common field worms (*Allolobophora caliginosa*), and their relatives. These creatures have linear digestive tracts enclosed within a long muscular tube divided into segments that can expand or contract independently. This anatomy enables segmented worms to burrow efficiently as they ingest soil at their front ends and simply pass it through to their anuses, absorbing nutrients from the ingested matter as it passes through their bodies. The 3200 or so known species of earthworms feed primarily on decomposing plant debris but may also consume small plant roots and soil organisms. Earthworms are found in numbers up to hundreds of thousands per acre or hectare of moist, rich soil and together may produce as much as 18 tons per acre (41 metric tons per hectare) per year of nutrient-rich manure (worm castings). Earthworms have long been regarded almost as saints of the garden for their roles in aerating the soil, improving drainage, mixing soil layers, and converting plant and microbial debris into worm castings. Of earthworms, Charles Darwin is reported to have said, "It may be doubted whether there are many other animals which have played so important a part in the history of the world, as have these lowly organised creatures." Aristotle called them the "intestines of the earth." The major natural enemies of earthworms include passerine birds, burrowing mammals such as moles, and a variety of nematodes, fungi, and bacteria.

Nematodes

Also known as roundworms or eelworms, nematodes are very thin, often quite active, smooth worms with pointed ends (plate 49). Soil-dwelling nematodes are usually microscopic or near-microscopic, with diameters of only a few microns (1 micron = 1/25,000th inch) and lengths of usually less than 1/25th inch (1 mm)—although some pregnant female nematodes may swell into much larger rotund cysts readily visible to the eye.

Despite their tiny sizes, nematodes have great influence in garden ecology, rivaling insects, bacteria, and fungi in importance under certain conditions. Plant losses due to nematode damage are estimated to range from less than 5 percent in cool, dry regions with harsh winters to greater than 25 percent in tropical regions with sandy soils. The most significant groups of plant-damaging nematodes are the root-knot nematodes (*Meloidogyne* and relatives) and cyst-forming nematodes (*Heterodera*), but there are a great variety of others. Soybean cyst nematodes (*Heterodera glycine*) are one of the major pests of soybeans in North America. Sugar beet cyst nematodes (*Heterodera schachtii*) have almost eliminated the growing of beets in some otherwise well-suited areas in North America and Europe. Nematodes are one of the greatest headaches of pepper and tomato growers in California and along the Atlantic Coastal Plain from Florida to New England. They are also serious pests of many other ornamental and food plants in Florida, Hawaii, and California.

Many nematodes are also parasites of mammals. When you worm a cat, dog, or horse, you're attempting to expel or kill parasitic nematodes—often picked up from soil-borne cysts or eggs deposited originally in feces from an infested animal (a good reason not to use cat or dog manure in a compost pile). Most soil-dwelling nematodes, however, are major predators or parasites of insects and other arthropods, other invertebrates such as rotifers, or even other nematodes. Many nematodes also prey heavily on protozoans, yeasts, molds, and bacteria. Increasingly, commercially raised parasitic or predatory nematodes are being used by environmentally conscious gardeners and farmers for biological control of insects such as grasshoppers and Japanese beetles (see chapter 6).

The body architecture of a nematode consists basically of two side-by-side tubular cavities, one for reproduction and one for digestion, within a long, narrow, muscular body wall. Nematodes usually feed by piercing larger victims with a retractable, hollow needle and sucking out their body fluids; by sucking in small bacteria, protozoa, bits of organic debris, or other objects small enough to fit in their mouths; or by pressing prehensile lips against a victim and boring their way in with sharp, miniature teeth or rasps, and then ingesting fluids or chunks of tissues. Some nematodes are strictly predatory, living their entire lives free in soil or water and moving from place to place in search of prey. Others are ectoparasites (ecto- meaning "outside"), attaching themselves to hosts but not entirely entering their bodies. Females of the major plant-parasitic cyst and gall nematodes partially embed themselves in a host and spend their lives in one place, but release their eggs or offspring to the outside environment. Other nematodes burrow entirely into the tissues of hosts and become endoparasites (endo- meaning "inside"). Nematodes also damage hosts by acting as vectors of various pathogenic fungi, bacteria, and viruses. Indeed, many plants genetically resistant to various soil-borne diseases effectively lose that resistance when planted in soil co-infested by certain nematodes and disease organisms. Some plant diseases known to be spread or greatly aggravated by nematodes include *Fusarium* and *Verticillium* fungal wilts; bacterial galls, rots, and wilts; and viral diseases such as ringspots, mosaics, tobacco rattle, pea early browning, and fanleaf of grapes.

Nematodes are found worldwide in virtually all soils and fresh waters, but are especially abundant and significant in coarse-textured soils of the tropics and subtropics that do not freeze. They can also be important in greenhouses if they become established. Nematodes are motile, but they do not travel very far nor very fast, and are often dispersed by human activities rather than by spreading on their own. Most nematodes enter hosts as they make contact with or ingest infested soil or water. Damage to plants is generally first seen as wilting, stunting, and yellowing and may be confused with nutrient deficiencies or fungal or bacterial diseases. Closer examination may reveal cysts or galls on roots, stems, or even seeds; rotting lesions of various kinds; and misshapen and truncated root systems. However, it is often difficult to definitively diagnose a nematode infestation without the help of a professional.

Other Garden Invertebrates

Quite a number of less familiar invertebrates can be numerous and widespread in garden soils. However, the ecological roles of these animals are not well known. But like the symphylans, they may not be fully appreciated simply because they have not been as well studied as more conspicuous organisms. The rotifers and nematomorphs, in particular, may merit more careful investigation because of their wide distribution and large numbers.

Tardigrades, or water bears, are near-microscopic, pill-shaped creatures encased in jointed plates of hard cuticle. They have four pairs of long-clawed legs with which they pull themselves along in water films. For the most part, tardigrades are plant parasites with piercing-sucking mouthparts, but their significance in damaging garden plants is little known, and some tardigrades are predators of nematodes and other small soil-dwelling invertebrates. Although primarily found in wet environments, water bears are capable of surviving prolonged drought or severe cold by entering a state of deep dormancy.

Hairworms, or nematomorphs, are common and widely distributed, near-microscopic creatures resembling extremely thin nematodes, except they lack mouthparts and excretory organs. Hairworms instead absorb nutrients through their skin. Juveniles live primarily as internal parasites of insects, other arthropods, and mollusks such as snails and slugs, but adults are free-living in water and wet soils. Nematomorph larvae are usually ingested by insects and other potential hosts as they drink water, and, although they rarely kill their hosts outright, they quite commonly cause reproductive sterility so that their impact on arthropod populations may be significant.

Rotifers, gastrotrichs, kinorhynchs, and acanthocephs are sometimes considered by invertebrate zoologists as architectural variants of nematodes, although most current taxonomic schemes assign them their own phyla. These soil-dwelling, near-microscopic organisms are often fantastic in appearance: rotifers are named for whorls of long cilia that surround their large mouths and whip about, giving the appearance of whirling or rotating spoked wheels. Most of these organisms are predaceous or parasitic on other animals, protozoa, or bacteria.

Onychophores are bizarre-looking creatures with heads resembling those of snails, bodies like shortened nightcrawlers up to 4 inches (10 cm) long, and rows of stubby legs somewhat arranged like those of a stylized centipede or caterpillar cartoon character. These little-studied predators live primarily in tropical and subtropical areas, where they dwell beneath leaves, rocks, and other larger debris and hunt smaller insects and other arthropods.

Flatworms are mostly internal parasites (flukes and tapeworms) of higher organisms including fish, mollusks, and mammals. However, there are free-living, soil-dwelling forms such as planarians (class Turbellaria) that prey on other small soil organisms including nematodes and rotifers, and some flatworms are detritivores. In parts of the world where fresh human manure (night soil) is used as fertilizer in gardens, tapeworms (class Cestoda) and flukes (class Trematoda) may pass from person to person via soil-contaminated vegetables and fruits.

Protists

The kingdom Protista includes a wide range of eukaryotic (containing nuclei in their cells) organisms that have cells similar to those of animals or plants but that lack differentiated tissues. Most protists are single celled. Many are microscopic individually, although they may cluster together in great numbers to form readily visible colonies—up to hundreds of miles across in the case of some sea-going protists. Of all the kingdoms of organisms, Protista is the most loosely defined; it includes organisms that at various times have been classified as plants, animals, or even fungi. However, most modern biologists consider protists to include the photosynthetic algae and diatoms as well as nonphotosynthetic, single-celled organisms generally known as protozoans. Many of these organisms are important in the ecology of soils and natural waters—and thus of gardens.

Diatoms and Algae

Diatoms, true algae, and their relatives are photosynthetic organisms that produce chlorophyll and other pigments, as do higher plants. They differ from plants primarily in having simple, undifferentiated tissues, as well as in their anatomical structures.

Diatoms have hard, essentially crystalline, transparent outer shells made of silica (a material similar to glass). They are very abundant in both salt and fresh water. As plankton, diatoms may be responsible for up to 20–25 percent of photosynthesis on Earth and are major food sources for larger aquatic and marine animals. Many diatoms also live in the surface layers of moist soils mixed among the true algae and are indistinguishable from them without microscopic examination. Some diatoms even grow on the surfaces of mollusks, arthropods, or turtles. We discuss diatomaceous earth, which is mined from ancient deposits of diatom shells, in chapter 6.

Marine algae such as the kelps and laver are large (up to 500 feet, 150 m long) multicellular organisms usually colored brown, gold, or red. Most soil-dwelling algae (phylum Chlorophyta), however, are microscopic, single-celled or loosely colonial, and bright green or yellow to yellowish green. So-called blue-green algae (or cyanobacteria; phylum Cyanophyta) physically and ecologically resemble green algae and historically have been grouped with them. However, cyanobacteria are actually prokaryotes (single-celled organisms lacking such subcellular structures as nuclei and mitochondria), and are now classified with the bacteria. But for the purposes of garden ecology, blue-green algae can be usefully considered in the historical fashion together with true algae.

Because they need light for photosynthesis, algae live mostly in the upper few inches of soil, where, under appropriate conditions, they may multiply to form greenish or blue-green crusts or blooms. The soil surface is commonly subject to wide fluctuations in temperature and moisture, however, and most soil-dwelling algae are capable of forming cysts or other dormant forms that can survive unfavorable environmental conditions. Soil-dwelling algae are found worldwide, even in polar regions and extremely harsh deserts. However, they seem most numerous and active under moist, well-illuminated conditions—such as in greenhouses or on temperate-zone soils in spring before leaves of plants significantly shade the soil. Populations of active soil algae may fluctuate widely from a few thousand to millions per ounce (hundreds to hundreds of thousands per gram), the latter during blooms. However, typical population counts run in the neighborhood of 25,000–100,000 per ounce (about 1000–4000 per gram) of soil, usually making algae much less numerous than bacteria, protozoa, fungi, or even nematodes.

Despite their small numbers under most conditions, soil-dwelling algae play significant ecological roles, especially where the soil remains relatively undisturbed (plate 50). Green and yellow-green algae secrete large quantities of gummy polysaccharides (complex carbohydrates) that are important in forming soil aggregates and maintaining soil structure. Blue-green algae are nitrogen-fixing organisms and may add up to 100 pounds per acre (110 kg per hectare) per year of organic nitrogen in pasture and woodland soils, a rate of addition comparable to some nitrogen-fixing legumes. Algae and diatoms are important food sources for nematodes, protozoa, rotifers, and other soil-dwellers that do not photosynthesize. There is also evidence that some algae release selectively toxic substances or antibiotics that affect populations of other soil microbes. Algae also form partnerships with fungi in the widespread symbiotic organisms known as lichens (see chapter 5).

Protozoans

Protozoans are a large and diverse group of single-celled organisms that are primarily nonphotosynthetic and usually, but not always, microscopic. Protozoans in the garden environment include the amorphous amoebas and their relatives; a wide variety of flagellates that move with whiplike appendages; ciliates, with beating, short, hairlike appendages; nonmotile coccidia, which are parasitic on other organisms; and euglenas, which can optionally live by photosynthesis or by ingesting animal or plant materials. Under most conditions flagellates are the most numerous of the protozoa in soils, but they have been much less studied than amoebas.

Protozoans require free water for locomotion, and most soil-dwelling protozoans subsist on a diet of bacteria, rotifers, nematodes, algae, and other small organisms. Therefore, protozoans are most abundant and diverse in moist soils rich in organic matter. Some tropical soil-dwelling amoebas may cause serious human disease if ingested, but most seem innocuous to humans as they ooze about and voraciously ingest other garden organisms, both harmful and beneficial. Some studies have shown that many species of amoebas are quite selective predators, however; thus, they may significantly impact the composition of populations of soil-dwelling bacteria, algae, and other organisms.

Slime Molds

Slime molds, or myxomycetes, are oddities of the biological world, having properties of single-celled protozoans (especially amoebas), during much of their life cycles, but greatly resembling brightly colored, soft-bodied fungi at other times. Taxonomists disagree whether to classify them as protists, fungi, or in a completely separate category.

Ecologically, slime molds behave in gardens generally as amoebas do, inconspicuously creeping about and ingesting other small organisms (including true amoebas) or bits of plant debris. Under conditions of nutrient deprivation, however, individual amoeboid slime mold cells send out chemical signals that attract other slime mold cells. These cells congregate to form large, motile masses often resembling brightly colored, gigantic amoebas, heaps of scrambled eggs, or other bizarre forms. These aggregations slowly creep about (reportedly, even over sleeping dogs!) until they settle down and form mushroom-like fruiting structures that release airborne spores (plate 51). Spores drift in air or water currents until they land in an appropriately moist and nutrient-rich spot, then germinate into tiny amoeboid creatures, thus restarting their life cycles. Slime molds are fascinating in appearance and behavior—it's a lucky gardener who gets to observe one of these natural oddities in action.

Fungi

Fungi (molds, yeasts, mushrooms, and their relatives) are very influential in the garden. Some are among the most aggressive plant pathogens and are responsible for most diseases of plants, whereas others are highly beneficial, even vital, symbiotic companions of plants (see chapter 5). Still others are nature's recyclers, playing major roles in natural nutrient cycles (see chapter 3).

Historically, fungi were classified with plants, although recent studies reveal that they are actually more closely related to animals. Like plants and animals, fungi are eukaryotic, but they are distinctive enough to be assigned their own biological kingdom in modern classification schemes. Like plants, they have cell walls but their very tough, carbohydrate-protein walls are quite unlike those of plants. Like animals, fungi are not photosynthetic, and they live by absorptive nutrition (digesting organic materials with secreted enzymes, then absorbing the dissolved nutrients). However, fungal cells often have the unique feature of containing many independent, genetically distinct nuclei, which confers considerable genetic complexity and flexibility.

Most fungi grow as hollow, often highly branched, tubules (hyphae) containing discontinuous stretches of gelatinous living tissues (plate 52).

Hyphae often grow and interweave into a dense, cottony mass called a mycelium, which has been fancifully described as "a tangled pile of near-microscopic garden hoses filled with amoebas." Yeasts do not grow as hyphae, but rather as thick-walled, single-celled spheres or ovoids that reproduce by budding off daughter cells. Some fungi, especially parasites of plants and animals, can change reversibly back and forth between yeast and hyphal forms, a capability known as dimorphism (from Greek for "two shapes").

Fungi reproduce in a bewildering variety of ways. Some produce prodigious numbers of diverse spores that drift or float over great distances and into the tiniest crevices. Fungi may also generate dormant vegetative structures, or propagules, that are long-lived and resistant to extreme environmental conditions. Many fungi can reproduce by fragmentation of living colonies. Under favorable conditions, fungal spores or vegetative propagules may eventually germinate to start new hyphae, which in turn grow into new fungal colonies. A fungal colony may spread for great distances underground or in plant tissues and come to the gardener's attention only when it sends up spore-forming reproductive (fruiting) structures, such as mushrooms (plate 53). In fact, at the time of this writing, the world's largest organism is arguably a colony (actually a genetically homogeneous individual apparently from a single spore) of honey mushrooms (*Armillaria ostoyae*) that occupies 2200 acres (880 hectares) of woodland soil in eastern Oregon. If it could be dug up intact and separated from the soil, this fungus supposedly would weigh thousands of tons.

Fungi play critical ecological roles in mobilizing and recycling nutrients (especially carbon, phosphorus, and metals such as zinc, iron, and copper); digesting complex molecules such as cellulose and lignin in dead plant tissues (plate 54); and forming beneficial symbiotic associations with plant roots (mycorrhizae). Other significant roles include helping to form secondary soil particle structures; parasitizing insects, nematodes, and other animals; and, most conspicuously to gardeners, acting as the dominant group of plant disease-causing organisms (see chapter 5).

A single ounce (28.4 g) of garden soil may contain several million viable fungal spores and propagules as well as considerable amounts of vegetative fungal tissue. As much as 50 to 80 percent of the entire living biomass (which also includes insects, worms, bacteria, and plant roots) in a garden soil sample may be fungal tissue. Fungi are especially prominent in dry, cold, or acidic soils or in those high in carbon-rich but nitrogen-poor organic materials such as sawdust, straw, fallen leaves, or dead woody plants. Other than some yeasts, however, fungi are strictly aerobic (they require oxygen) and fare poorly in flooded or waterlogged soils, where bacteria become the overwhelmingly dominant microflora.

The taxonomy and nomenclature of fungi are so arcane and complex—and undergo such constant revision—that even specialists in the study of fungi (mycologists) can become bewildered. Generally, however, mycologists classify fungi based on their modes of reproduction (especially whether or not they form motile spores) and on the structures of their reproductive parts, the details of which are beyond the scope of this book.

Among the major groups of fungi important in garden ecology are the imperfect fungi (also known as Fungi Imperfecti, deuteromycetes, phylum Deuteromycota, and other terms). These fungi are not known to reproduce sexually, but usually form great powdery masses of often brightly colored asexual spores by a cloning process, or they form no spores at all and reproduce only by dispersal of vegetative propagules. Nevertheless, the 15,000 or so species of imperfect fungi are probably the most abundant and ubiquitous of all fungi. They include most detritivores, such as multitudinous species of *Aspergillus*, *Penicillium*, *Cladosporium*, and *Trichothecium*, which are found in virtually every sample of soil worldwide. Some soil-dwelling imperfect fungi, such as species of *Trichoderma* and *Trichophyton*, live in the soil closely surrounding plant roots, where they feed on nutrients leaking from roots and help to defend plants from attack by pathogenic fungi and bacteria. Species of *Trichoderma* are now commercially available for use as natural pesticides to protect seeds and seedlings of garden plants (see chapter 6). However, some of the most destructive plant pathogenic fungi are also imperfects.

Approximately 20,000 species of sac fungi, or ascomycetes (phylum Ascomycota), are known. These include most yeasts, which live predominantly on the surfaces of fruits, stems, and leaves, where they consume sugar-rich exudates. Other sac fungi are parasites of animals. Truffles, morels, earth tongues, many saprophytic molds, most fungi that form lichens, and the majority of plant-pathogenic fungi are ascomycetes. Club fungi, or basidiomycetes (phylum Basidiomycota), comprise about 15,000 known species, including the great majority of mushrooms, puffballs, earth stars, jelly and ear fungi, most ectomycorrhizal fungi (see chapter 5), most wood-rotting fungi (including the polypores or bracken fungi), and several major groups of plant pathogens. Zygomycetes (phylum Zygomycota) include the virtually ubiquitous, fast-growing, massively spore-forming genera *Rhizopus* and *Mucor*—the weeds of the fungus world, found world-wide on moldy bread. Other zygomycetes include important pathogens of insects such as *Entomophthora* ("insect eater") and many soil-dwelling predators and parasites of nematodes, amoebas, and protozoa.

The so-called lower fungi, also traditionally known as phycomycetes, live primarily in aquatic environments, in damp soils, or on plant surfaces in regions where rain or dew is frequent. The largest group of phycomycetes is the egg fungi, or oomycetes (phylum Oomycota), a heterogeneous group. One subgroup, the water molds (order Saprolegniales), are primarily aquatic detritivores, being especially numerous in heavily polluted waters. Some oomycetes are parasites of fish or other aquatic animals or, more rarely, plant roots in wet soils. Other oomycetes, the order Peronsporales, are mostly terrestrial and parasitic on higher plants and include notorious plant pathogens that cause diseases known as downy mildews. Other phycomycetes, the chytrids (phylum Chytridiomycota) and hyphochytrids (phylum Hyphochytridiomycota), are mostly aquatic, although some live in damp soils. These groups are primarily parasites of algae, protozoa, and other small organisms, although some live on dead plant or animal tissues and a few are parasitic on higher plants. Plasmodiophoromycetes (try saying that quickly ten times!) are mostly parasites of higher plants, burrowing into plant cells, then losing their cell walls to become amoeboid organisms that cause a cancerlike growth of plant tissues.

Actinomycetes

According to their genetic and biochemical traits, actinomycetes are a sub-group (most commonly the class Actinomycetales) of the bacteria (treated hereafter). Because they have distinctive structures and ecological roles, however, they are often considered separately by ecologists and agricultural scientists. Actinomycetes (also called bacterial fungi) form slow-growing colonies of tangled, leathery, highly branched tubules with elaborate, erect, spore-producing structures, indeed resembling some fungi in form.

Actinomycetes are important in garden ecology and are abundant in soil—with hundreds of thousands to several million colony fragments or viable spores per ounce (thousands to hundreds of thousands per gram) of fertile garden soil. They are major producers of toxic organic substances (antibiotics), which may aid them in competing with faster-growing bacteria and fungi. Some actinomycete-produced antibiotics, such as streptomycin, tetracycline, erythromycin, neomycin, nystatin, chloramphenicol, and clindamycin, have been adopted for use in human medicine. Many actinomycetes secrete powerful enzymes that digest bacterial cell walls, causing the bacterial cells to rupture. Chemical warfare by actinomycetes may be especially significant in suppressing soil-dwelling fungal and bacterial pathogens. Except in the uppermost surface layers of soil, where algae become significant, actinomycetes may play the primary role in creating secondary soil structures such as crumbs, blocks, and prisms that prevent soil from collapsing under its own weight. Actinomycetes are predominantly aerobic detritivores that are especially fond of starch and chitin, the major structural material in the exoskeletons of arthropods and in the cell walls of many fungi. There are also a few significant plant disease-causing, parasitic actinomycetes. And, trivial perhaps, but interesting nevertheless, actinomycetes synthesize geosmin, the substance largely responsible for making soil smell "earthy."

Actinomycetes can tolerate severe drought and tend to be most abundant in warmer, drier soils of neutral or alkaline pH. They are most significant in desert, steppe, or prairie soils—as may be indicated by the potent earthy aroma of many such soils. In cold, acidic, or waterlogged soils, actinomycetes are much less common, although they are almost never entirely absent. Major groups of soil-dwelling actinomycetes include species of *Streptomyces*, which are famous for producing medically important antibiotics; nitrogen-fixing *Frankia* species, which form symbiotic associations with the roots of several nonleguminous woody plants; *Micromonospora* and *Thermoactinomyces* species, which are so abundant in vigorously decomposing (heating) compost that they often seem to form a white dust; and *Arthrobacter* species, one of the most abundant microorganisms in very dry soils.

Bacteria

Bacteria are the most numerous and most diverse organisms on Earth. They have been found in almost every habitat, no matter how extreme. In fact, bacteria are so diverse in their biochemical and genetic characteristics that modern biologists have reclassified them into at least two separate kingdoms: Eubacteria (most significant in gardens) and Archaea (found mostly in extreme environments). However, the vast majority of bacteria on Earth live in one habitat—the soil. One ounce (28.4 g) of fertile garden soil may contain several billion living bacteria. Some of the most critical biochemical processes in the garden environment are carried out mainly by bacteria. Although a gardener will never see a single individual bacterium without the aid of a high-powered microscope, their influence, both for good and ill, cannot be ignored.

Bacteria are prokaryotes that range from 1/50,000th to 1/2500th inch (0.5–10 microns) in maximum dimensions (although there is a recent report of a giant bacterium barely visible to the naked eye found in the ocean off the southwest coast of Africa). In form bacteria range from helplessly drifting spheres to elongated corkscrews that bore their way at high speeds through liquid media. Bacteria are generally distinguished and classified primarily by their biochemical processes and physiological characteristics, of which they have a bewildering variety—many found in no other group of organisms.

Many bacteria break down sugars, fats, and other natural organic materials in the presence of oxygen to produce energy, water, and carbon dioxide, much as do plants, animals, and fungi. However, some bacteria are also capable of metabolizing seemingly indigestible substances such as petroleum, motor oil, and kerosene. Bacteria also break down pesticides, synthetic fabrics, dyes, glues, cleansers, antibiotic drugs, and even some plastics.

Many bacteria are oxygen-producing photosynthesizers. Others carry out photosynthesis that does not produce oxygen, trapping light energy to convert carbon dioxide to sugars, but using hydrogen sulfide, carbon monoxide, or other chemical substances in place of water to create the sugars, and generating solid sulfur or other materials instead of oxygen as by-products. Bacteria such as *Clostridium botulinum*, of food poisoning notoriety, live by anaerobic fermentation, breaking down sugars in the absence of oxygen to various acids and/or alcohols plus energy. Still other soil-dwelling bacteria carry out anaerobic respiration, in which they metabolize sugars and other organic materials with nitrate, sulfate, or other inorganic substances rather than with oxygen. No other organisms are capable of this process, which is very important in the cycling of nitrogen, sulfur, and other nutrients (see chapter 3). Strangest of all are the chemolithotrophic bacteria, which power their life processes by inorganic chemical reactions, such as the rusting of iron or electrical battery-like reactions between minerals. Chemolithotrophic bacteria are neither rare nor unusual in nature, however, and many natural chemical processes in soil, waters, and rock depend on their activities. Some mineral ore deposits are thought to be the fossilized remains of such bacteria.

Bacteria may reproduce extremely quickly. Under ideal conditions, some species divide every fifteen minutes—leading to as many as ninety-six generations in a single day. Were it not for predators of bacteria such as protozoa and nematodes, viral diseases (yes, even bacteria get sick), or nutrient limitations, in theory, over two days a single bacterium of a faster-growing species could produce descendants equal in mass to the entire planet!

Bacteria also have many distinctive genetic characteristics. For one thing, they are often quite unstable genetically and are prone to frequent mutations. Many also freely exchange genetic material with one another or pick up and incorporate snippets of genetic material from other types of organisms. Some can even use stray scraps of DNA found in the environment. An important consequence of this genetic promiscuity is that bacteria can pass new genetic traits or mutations not only to their descendants, but also to their neighbors. (Humans did not invent genetic engineering.) Their combination of huge populations, extremely fast reproduction, frequent genetic mutation, and genetic promiscuity has a very important ecological consequence: bacteria are able to adapt quickly to virtually any environmental influence via natural selection. This is one major reason why use of antibiotics will never eradicate pathogenic bacteria—the microbes adapt and become resistant all too quickly.

Bacteria predominate over other soil microbes in wet, nutrient-rich environments and are overwhelmingly dominant in stagnant water and in waterlogged or compacted soils. In such low-oxygen environments, bacteria may comprise over 99 percent of the microflora. Anaerobic bacteria make soluble large amounts of heavy metals. They also produce acids such as sulfuric and acetic acids; release gases such as methane, hydrogen sulfide, ammonia, and carbon monoxide; and synthesize many other substances toxic to most plants and other organisms. In fact, anaerobic bacteria are probably the major cause of the harmful effects of waterlogged soils to other organisms.

Eubacteria, or true bacteria, have traditionally been divided into two groups, Gram-positive and Gram-negative. These labels refer to whether or not they are capable of adsorbing a purple-black dye developed by the nineteenth-century Danish microbiologist Christian Gram. The dye detects fundamental differences in cell wall structure and chemical composition between the two groups, which also differ in significant physiological and ecological characteristics. The great majority of soil-dwelling bacteria are Gram-positive, including the large genera *Arthrobacter*, *Bacillus*, *Clostridium*, and *Streptococcus*. Gram-positive bacteria tend to be mechanically tougher than Gram-negative bacteria and more resistant to environmental stresses such as heat, cold, and desiccation. However, virtually no Gram-positive bacteria are motile and most are quite sensitive to poisoning by antibiotics. Relatively few Gram-positive bacteria are pathogens or symbionts of either plants or animals, but there are some major exceptions, especially in the genera *Frankia*, *Streptococcus*, *Staphylococcus*, and *Corynebacterium* (which causes diphtheria, as well as a number of plant ailments). However, Gram-positive bacteria such as *Bacillus* and *Clostridium* are capable of forming extremely long-lived, heat-resistant dormant spores—a fact of great consequence to home canning of vegetables because botulism is caused by *Clostridium botulinum* from the soil. Spores of *Bacillus anthracis* have achieved notoriety as agents for spreading the deadly disease anthrax. Other spore-forming species of *Bacillus* are used for biological control of insect pests (see chapter 6).

Gram-negative eubacteria (plate 55) are a much more diverse group than are Gram-positives, being united solely by their not having Gram-positive cell walls. Gram-negatives tend to be more susceptible to injury by heat, cold, desiccation, and other physical forces than are Gram-positives, but they are generally more resistant to antibiotics and other toxic chemical substances, being shielded by water-repellent layers in their cell walls. Most plant-pathogenic (and, for that matter, animal-pathogenic) bacteria are Gram-negatives. Although less numerous overall than are Gram-positives in the garden environment, Gram-negatives are more important in mineral nutrient cycles and the degradation of complex organic molecules.

Gram-negative bacteria that are important in garden ecology include members of the soil-dwelling genus *Pseudomonas* that are especially famous—or infamous—for their ability to metabolize virtually any low-molecular-weight organic substances, including most pesticides, antibiotics, and industrial chemicals. *Rhizobium* species are important in nitrogen-fixing symbioses with plants, and bright-yellow *Xanthomonas* are notorious plant pathogens. The corkscrew-shaped bacterial torpedo *Bdellovibrio* kills other bacteria by ramming them at high speed, equivalent on a human scale to approximately 400 miles/hour (650 km/hour)! *Thiobacillus*, *Nitrosomonas*, and *Desulfovibrio* are important in nutrient cycles of substances such as nitrogen, sulfur, iron, and copper. Nitrogen-fixing, photosynthetic, aerobic cyanobacteria such as *Nostoc*, *Anabaena*, and *Chroococcus* (also known as blue-green algae) help maintain fertility of many woodland, wetland, and pasture soils. Predatory bacteria such as *Cytophaga* help control populations of other bacteria. Because it is capable of carrying genetic material from one plant to another, the crown gall–causing plant pathogen *Agrobacterium* has been used extensively in genetic engineering of plants. Free-living nitrogen-fixing *Azospirillum* and *Azotobacter* are important in maintaining the fertility of soils in Asia, where some fields have been farmed for millennia without the addition of any fertilizers.

The mycobacteria, rickettsias, chlamydias, and mycoplasmas are groups usually considered separately from either the Gram-positive or Gram-negative eubacteria. Mycobacteria, rickettsias, and chlamydias are generally parasites of animals—some causing important diseases such as tuberculosis, Rocky Mountain spotted fever, and leprosy—but generally are of little known significance in garden ecology. In contrast, mycoplasmas and mycoplasma-like organisms (MLOs), which are extremely flexible, amoeboid bacteria that lack cell walls, have become recognized in recent years as causative agents of a number of previously mysterious soil-borne plant diseases. Some MLO contaminations of soil effectively prevent raising susceptible plant species for decades after afflicted plants have been removed and destroyed. One example is the so-called stale soil syndrome of many declining rose family fruit orchards.

Archaea, Viruses, and Prions

Most of the bacteria-like organisms that inhabit extreme environments—abyssal ocean depths, boiling sulfuric acid springs, saturated salt brines, volcanic vents on the ocean floor that belch hot gases and water, and rocks thousands of feet below the surface of the Earth—are archaea (also called archaebacteria). These resemble bacteria in gross form and function but have such different biochemical and genetic characteristics that they are now classified by many microbiologists in a separate kingdom (Archaea). Few of these organisms have been reported from more normal environments such as gardens, but that may be simply because not many searches have been made for them there.

The largest viruses are about the size of the smallest bacteria and are barely visible under the most powerful light microscopes. Most viruses are so tiny that they can only be seen using electron microscopes. Viruses are basically bags or canisters made of protein and/or fats that contain genetic material (DNA or RNA). They cannot metabolize or reproduce outside of the cells of a living host, and they never grow, but only duplicate themselves. Viruses are also incapable of any independent motion and must be transported to and introduced into a host by another organism, a vector. Even so, viruses cause many serious diseases of garden animals, plants, protozoa, fungi, and bacteria. When introduced into the soil outside of a host or vector organism, most viruses are inactivated by being adsorbed by soil particles or are digested into fragments by enzymes secreted by bacteria and fungi. However, it is likely that bacteria-infecting viruses, bacteriophages (meaning "bacteria eaters"), attack many bacteria in the soil, although this aspect of soil ecology has been little examined. Soil-borne fungus-infecting viruses are major problems for commercial mushroom growers, at times causing massive destruction of mushroom beds. In the oceans, bacteriophages are believed to be the agents primarily responsible for controlling populations of marine bacteria.

For many years scientists doubted the very existence of prions, which are mere bits of protein that seemingly can infect higher organisms and multiply inside them, doing serious damage in the process. Prions are very stable to treatment with heat, cold, desiccation, and a broad range of chemicals, so they may persist intact in the environment and remain infective for long periods if released from hosts. It is now generally accepted that the incurable diseases of bovine spongiform encephalopathy (BSE, or mad cow disease), scrapie of sheep, and Creutzfeldt-Jakob syndrome of humans are caused by prions. This has led to controversy among organic gardeners over whether to continue the use of bloodmeal and bonemeal as fertilizers, as there is concern that these materials could transmit mad cow disease to humans. Some ill-defined, although apparently infectious, plant diseases such as peach tree decline and sick soil of old orchards may also be due to prions as well as to mycoplasmas or MLOs. There is no known cure for these diseases, and all that can be done at present is to destroy the sick plants and fallow infested soil or grow plants that are not susceptible.

☙❧

The garden ecosystem is crowded with a large and diverse cast of characters, many of which have great influence on the well-being of plants and on each other. A shopper for a new home wisely considers the neighborhood and the neighbors when buying or building a new home. So also should a wise gardener try to ensure that plants will get along with their neighbors, both friendly and hostile, and take measures for civic improvement, if necessary.

3

The Garden Environment

Plants and other garden organisms do not exist in a vacuum. They dwell and function in a complex physical setting comprised of a multitude of nonliving components and forces that interact with one another in diverse ways. These abiotic factors powerfully influence living organisms and, in turn, are acted on by living creatures. Some of the most influential environmental factors affecting garden organisms are sunlight, air, water, and soil.

Sunlight

Almost all life on Earth depends directly or indirectly on the sun, and plants depend on the sun more directly than do other organisms. Although the Earth intercepts only about one six-quadrillionth of the sun's total radiative energy output, that fraction is enough to maintain the surface of the Earth at an average temperature of about 57°F (14°C). This energy also evaporates billions of gallons of water daily from the Earth's surface, generates powerful air and water currents that redistribute water and heat across the planet, and drives the critical plant life processes of photosynthesis and transpiration.

Sunlight must run a gauntlet of clouds and dust in the Earth's atmosphere, and much sunlight is scattered or absorbed on its way to the Earth's surface. Much scattered and absorbed sunlight is re-emitted from clouds as diffuse light, which is often of longer wavelengths than that of incoming light. Due to this diffusion of sunlight, shadows on Earth are gray rather than black, and both plants and gardeners can still get sunburned on cloudy days.

Sunlight that does get through the gauntlet is absorbed or reflected by water, plants, and soil. Absorbed light may be converted into heat, used to drive photochemical processes, or re-emitted as longer wavelength energy. Energy re-emitted from plants and soil, usually of infrared or longer wavelength, may escape back into outer space, leading to radiant cooling of the Earth, as occurs on clear, frosty autumn and spring nights. It may also be trapped by water vapor, especially in clouds, or by greenhouse gases such as carbon dioxide, leading to warming of the Earth. On a smaller scale, gardeners may also trap the Earth's radiant energy and warm their plants by using blankets, floating row covers, hotcaps, glass or plastic sheets, or other devices.

The amount of sunlight that is reflected varies from only a few percent of incident light from surfaces such as damp, dark-colored soil or a forest canopy to a very high percentage from surfaces such as snow or light-colored sand or rocks. Some reflected light may significantly increase photosynthesis by striking and being absorbed by otherwise shaded stems and leaves. Gardeners can take advantage of reflected light by putting white plastic or metallic foil beneath plants to reflect light upward to the undersides of leaves.

When the sun is high in the sky, sunlight on Earth appears white due to a fairly even intensity across all wavelengths of the visible spectrum. When the sun is closer to the horizon, such as at sunrise and sunset or during winter, refraction of sunlight by suspended particles makes the light reaching the ground increasingly yellow-orange in color. This phenomenon occurs because shorter wavelengths of light (ultraviolet, violet, blue) are absorbed and scattered in the atmosphere more than are longer wavelengths (orange, red, infrared). Plants may photosynthesize more efficiently in summer or near noon not only because of greater light intensity, but also because of the greater photosynthetic effectiveness of blue light as compared with that of yellow-orange light.

As discussed in chapter 4, day length has a dramatic effect on growth and development of many plants. The magnitude of changes in day length is strongly influenced by latitude. On the equator there are twelve hours of light and darkness year-round, whereas at latitudes greater than 67° day length fluctuates from twenty-four hours per day of light (summer) to twenty-four hours of darkness (winter) and back again. Furthermore, the rate of change of day length through the year is quite uneven, changing very little over the whole Earth near the summer and winter solstices (approximately June 21 and December 22), but very rapidly near the equinoxes (about March 21 and September 21), especially at high latitudes. Another seasonal effect on day length is that all the days in autumn are shorter than twelve hours, whereas all those in spring are longer than twelve hours—a rather confusing but significant fact for raising day-length-sensitive plants (plate 56). The upshot of all this day-length variation is that gardeners may have to match garden plants and planting times not only with their latitude and consequent day length, but also with direction and rate of change in day length.

Air

The atmosphere is the ultimate source of carbon, oxygen, and nitrogen for plants. It transports water and heat from place to place. The atmosphere modifies solar irradiation (for both good and ill). It moves soil particles, salt spray, and plant debris from one place to another. It transfers wind-blown pollen from plant to plant and makes it possible for pollinating insects and birds to fly. The atmosphere distributes seeds and fruits including, unfortunately, some of those of the most noxious weeds. In the form of weather, the atmosphere can nourish, break, abrade, and uproot plants. It may deliver or remove toxic air pollutants and acid rain. It may dehydrate plants and soil or envelop them in dew and fog. And it strikes with lightning, which is a surprisingly common cause of plant injury and death in some areas, but also a significant source of fixed nitrogen available for plant nutrition.

The lower atmosphere is a mixture of gases averaging 78–80 percent nitrogen, 18–20 percent oxygen, 0.9 percent argon, 0.04 percent carbon dioxide, highly variable amounts of water vapor (0.02–4 percent), and variable amounts of sulfur and nitrogen oxides, helium, ozone, carbon monoxide, and suspended particles (for example, dust, pollen, salt crystals, smoke). However, the composition of the air close to the ground and in confined spaces may deviate substantially from that of the open sky. Among closely spaced plants that are vigorously photosynthesizing, carbon dioxide may become all but exhausted, whereas oxygen (released as a by-product of photosynthesis) may rise significantly above 20 percent, and humidity also may be very high. In some conifer forests a bluish haze is often apparent in the air due to large amounts of volatile hydrocarbons vaporizing from the plants (hence the "Blue Ridge" Mountains). In air within the soil, the oxygen content may drop as low as 12–15 percent, whereas carbon dioxide may rise to as much as 10 percent and humidity may reach super-saturation.

Gaseous nitrogen, which is essentially chemically inert, mainly contributes to the physical properties of air such as viscosity, pressure, fluid flow, friction, and surface tension. Nitrogen-fixing microbes, lightning, and fertilizer factories, however, are capable of converting atmospheric nitrogen into forms available for plant nutrition—at the cost of a tremendous input of energy (see chapter 5, and later in this chapter). Argon is even more chemically inert than nitrogen and contributes little to the overall properties of the atmosphere.

Carbon dioxide in the air is the major source of carbon for photosynthesis, carbonate for mollusk shells, and, ultimately, limestone, petroleum, and coal. It is also an effective greenhouse gas, trapping and holding radiant energy within the atmosphere. At high concentrations, as found in many soils, carbon dioxide acts as a narcotic sedative for many higher organisms, and soil-dwellers must be specially adapted to keep from being put to sleep. Carbon dioxide originates ultimately from volcanic emissions, but over the eons most released CO_2 has been stored in the forms of plant tissues, soil humus, limestone, coal, or petroleum. However, in the past century or two, mass burning of wood and fossil fuels by humans has caused a pronounced increase in the carbon dioxide content of the atmosphere, which increase may have as yet uncertain but possibly dramatic impacts on the climate of the Earth.

Water vapor, although a small fraction of the total atmosphere, has critical roles in garden ecology. It is the source of moisture for most plants (in the form of precipitation) and acts as an efficient reservoir of heat energy, buffering the air against sudden changes in temperature (plate 57). Dry desert air chills quickly after sunset, whereas humid jungle air may keep folks sweating all night long. Sprinkling garden plants with water, even chilly water, is a fairly effective antifrost tactic; the water releases large amounts of energy as it cools and freezes, thus protecting the plants. Conversely, evaporating water chills other materials with which it is in contact. Well-watered plants can often survive far more severe heat than can desiccated ones, as unrestricted transpiration removes large amounts of heat from the plants. It's well known that a large shade tree by a home is worth several thousand BTUs of air-conditioning capacity as the tree cools the air around it by transpiration as well as by shading.

Sulfur and nitrogen oxides, produced mostly by volcanic activity and burning fossil fuels, are generally found only in traces but have significant impacts on garden plants. Both are highly reactive materials that combine with oxygen and water to form sulfuric and nitric acids, which are major components of acid rain. On the other hand, these compounds may also contribute available sulfur and nitrogen nutrients to plants.

Suspended particulate matter, such as smoke, dust, or volcanic ash, although comprising an extremely small proportion of the total mass and volume of the atmosphere, plays a role far out of proportion to its quantity. Even very small quantities of particulate matter can markedly reduce the transparency of the atmosphere to both incoming sunlight and outgoing infrared radiation (plate 58). In spring and autumn, radiation frosts commonly occur on clear nights; overcast or foggy nights (with much suspended matter in the air) are generally much warmer. An old-time, but effective, antifrost measure in many orchards was lighting smudge pots to produce dense clouds of smoke (this practice is now largely banned due to air pollution concerns). The 1991 eruption of Mount Pinatubo in the Philippines sent a giant ash cloud 22 miles (35 km) into the sky and reduced the surface temperature of the entire Earth by an estimated 1°F (0.6°C) for about 18 months, with measurable effects still apparent 7 years later. Both 1816, the year after the massive eruption of the Indonesian volcano Tambora, and 1884, the year after the volcano Krakatoa threw perhaps 1 cubic mile (5 cubic kilometers) of dust into the sky, were referred to as "the year without a summer," with frosts in July as far south as New York City. One cubic mile is a tiny amount compared to the approximately *two billion* cubic miles of the lower atmosphere throughout which this volcanic dust was distributed, but even this small proportion strongly influenced global weather. Dust is powerful stuff.

On a more positive note, atmospheric particulate matter is essential for precipitation. Without suspended particles on which water molecules can condense, no amount of chilling could enable atmospheric water vapor to aggregate into masses large enough to fall to Earth. Cloud seeding is an attempt to trigger rainfall by spraying fine particles into clouds rich with water vapor. Also, rainstorms frequently follow billowing clouds of smoke from volcanoes or large fires. The year after Krakatoa's eruption was not only cold but very wet worldwide. (On the bright side, 1884 was renowned for spectacular sunsets, as the brilliant hues of sunsets and sunrises are largely due to diffraction of sunlight by suspended atmospheric particles.)

At sea level the atmosphere exerts a pressure of 14.7 pounds per square inch (psi; 100 kPA); this gradually decreases to about 12.6 psi (86 kPA) at 5000 feet (1500 m) elevation and 4.4 psi (30 kPA) at the top of Mount Everest (29,035 feet, 8850 m). One consequence of the decreasing atmospheric pressure with increasing elevation is decreasing available oxygen and carbon dioxide. Plants growing on the Altiplano of Peru and Bolivia (elevation 11,000–13,000 feet, 3300–4000 m) or the Tibetan Plateau (15,000 feet, 4600 m) must adapt to oxygen and carbon dioxide availability only 50–60 percent of that at sea level—as well as to chronic cold, because the temperature of the atmosphere declines an average of approximately 3.5°F for each 1000 feet (6.5°C per 1000 m) increase in elevation.

Another attribute of the atmosphere is turbulence, of which the most obvious manifestation is wind (plate 59). Wind greatly accelerates the evaporation of water from plants and soil, which is a good news–bad news situation for plants. Without such evaporation, the vital process of transpiration, which draws water from the roots up to the leaves, could not proceed and plants would be unable to obtain nutrients from the soil. Diffusion of chemical substances in calm air is surprisingly slow, so without at least moderate atmospheric turbulence, plants can deplete the carbon dioxide immediately adjacent to their leaves and suffer growth limitation. Atmospheric turbulence also reduces excessive localized heating or chilling of the air and disperses concentrations of air pollutants. The dense smogs or fogs that often gather in valleys result when suspended materials accumulate in calm air layers (plate 60).

Excessive transpiration, however, can lead to wilting or even death from dehydration (plate 61). A breeze of as little as 5 miles/hour (8 km/hour) increases transpirational water loss 200 to 400 percent over that in still air. Plants such as tea (*Thea sinensis* or *Camellia sinensis*), rubber (*Hevea brasiliensis*), Norfolk pine (*Araucaria heterophylla*), and *Citrus* trees are so susceptible to windburn that it is difficult to raise them in areas of high winds or low humidity without substantial windbreaks. Wind also increases chilling of wet plants through evaporative cooling and aggravates winter-kill of unprotected perennials—evidence that winter-kill is actually a freeze-drying process, rather than simply a result of low temperatures.

Plate 1. In cold, exposed arctic and alpine environments, many plants, including purple saxifrage (*Saxifraga oppositifolia*), assume a cushion growth form.

Plate 2. Simple leaves are formed of a single leaf blade. In *Caladium*, red and purple pigments impart a distinctive look to the prominent veins and leaf surface.

Plate 3. Leaves in hobblebush (*Viburnum lantanoides*) are opposite, that is, borne in pairs along the stem. Also evident in this photograph are large, sterile flowers around the outside of each inflorescence that help attract pollinators.

Plate 4. Leaves of cardinal climber (*Ipomoea ×multifida*) have a long petiole and a distinctive, deeply lobed leaf blade.

Plate 5. In the leaves of grasses and related plants (monocots), the major veins run parallel to each other.

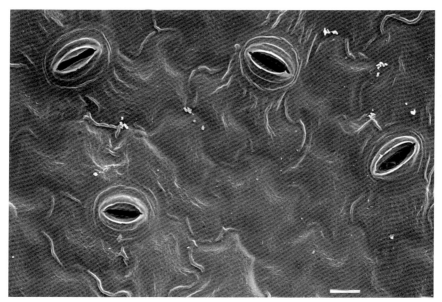

Plate 6. This scanning electron micrograph shows stomates in the epidermis of a geranium (*Pelargonium*) leaf. The jigsaw-puzzle-like shape of the epidermal cells helps maintain the integrity of the epidermis during leaf expansion (×750). Courtesy of Jeffrey M. Osborn

Plate 7. The primary function of most leaves is to capture light to carry out photosynthesis, but leaves in some species have other specialized functions. In Venus flytrap (*Dionaea muscipula*), the modified leaves spring shut to capture small invertebrates, such as insects, as a source of supplemental nutrients.

Plate 8. A primary function of most plant stems, including woody tree trunks, is to aid leaves in reaching light.

Plate 9. The fleshy stem tissue of this cactus carries out photosynthesis, stores water, bears protective spines, and supports flowers and fruits.

Plate 10. Strawberries (*Fragaria*) spread when stolons extend over the ground surface and put down roots.

Plate 11. Corn/maize (*Zea mays*) is characterized by prop roots that form at the base of the stem and provide support.

Plate 12. *Kalanchoe* propagates vegetatively when plantlets form in indentations along leaf edges. Adventitious roots can be seen on these plantlets.

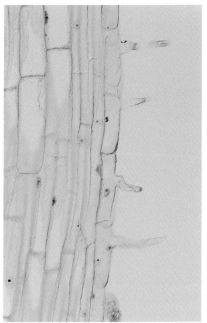

Plate 13. Water absorption in roots occurs through delicate root hairs formed in a narrow band a short distance back from the root tip. Courtesy of Jeffrey M. Osborn

Plate 14. In passion flowers (*Passiflora*), the male and female parts sit above a dramatic, filamentous corona. Courtesy of Anne Bergey

Plate 15. This Asian honeysuckle (*Lonicera japonica*) is widely planted for its sweet-smelling flowers. Only pollinators with relatively long tongues, such as hawkmoths (*Manduca*), are able to reach the nectar at the base of the narrow floral tube.

Plate 16. Cellulose is the chief chemical component of the plant cell wall, which makes up most of the mass of wood. This photograph shows a thin section through mulberry (*Morus*).

Plate 17. The radish (*Raphanus sativus*) in the foreground is an albino mutant that lacks functional chlorophyll throughout most of its tissue; its growth is significantly stunted, and it is unlikely to survive for long.

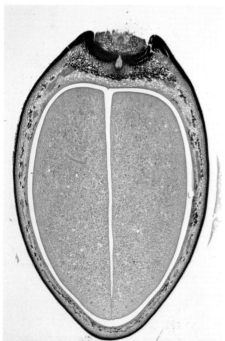

Plate 18. This thin section through a bean seed shows that most of its volume is taken up by the two cotyledons. Courtesy of Jeffrey M. Osborn

Plate 19. The outer fruit wall, which provides protection for the seeds within, varies considerably among fruits. Examples here include the relatively thin-skinned lingonberry (or mountain cranberry, *Vaccinium vitis-idaea*) and the hard-walled birdhouse gourd (*Lagenaria siceraria*). Photograph of birdhouse gourd courtesy of Kent Whealy, Seed Savers Exchange

Plate 20. In dicots such as this culti-vated *Citrus*, cell division at terminal meristems leads to an increase in height and branch length. Meristems also form in buds, which can be seen in the leaf axils of this plant. When these buds are released through pruning, herbivory, or other means, new branches grow out, causing increased spread.

Plate 21. This lemon grass (*Cymbopogon citratus*) was cut back, much as turfgrass is mowed. The younger, central leaves have since grown back, surrounded by the previously cut older leaves.

Plate 22. Cytokinins contribute to increased production of tissue, which results in the formation of galls. Thousands of different plants, including this species of *Rubus*, are stimulated to form galls by insects and other invertebrates. Courtesy of George Shinn

Plate 23. In seeking nectar from blue sage (*Salvia azurea*), this bumble bee (*Bombus* sp.) "cheats" by drilling through the side of the floral tube. Other bees enter the corolla tube; in the process, they pick up pollen, which may then be transferred to the next flower that they visit.

Plate 24. The sexes are separate in willows (*Salix*). In this close-up of a male willow catkin, the yellow pollen is strikingly evident.

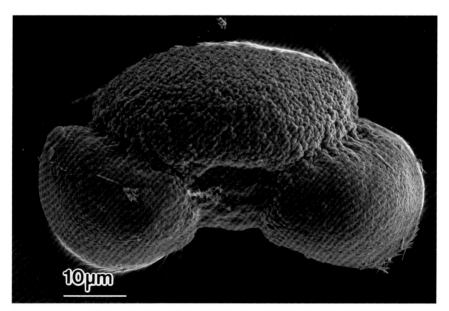

10μm

Plate 25. Pine pollen grains have characteristic air sacs, allowing them to be easily transported by wind (×1500). Courtesy of Jeffrey M. Osborn

Plate 26. In order for fertilization to occur, sperm must be transported through a pollen tube from the stigma, where pollen is deposited, to the ovary at the base of the style, shown here in this partially dissected lily (*Lilium*).

Plate 27. Although the flowers and leaves of the Indian paintbrush (*Castilleja linariaefolia*, left) and elephant head (*Pedicularis groenlandica*, right) do not overtly resemble each other, these two species are classified together in the snapdragon family (Scrophulariaceae) on the basis of all available evidence. Photograph of Indian paintbrush courtesy of Michael Ira Kelrick

Plate 28. Use of common names can be confusing. Each of these plants is called a primrose, although the plant on top is in the Primulaceae (primrose family) and the plant below it is in the Onagraceae (evening primrose family)

Plate 29. Squirrels may damage garden plants while burying nuts and acorns but propagate nut-bearing trees in the process. Courtesy of George Shinn

Plate 30. Dogs can be helpful in keeping other animals out of gardens, but they also love to dig and lie in newly turned soil.

Plate 31. Deer are major garden pests across North America and elsewhere in the world. Deer habituated to feeding in gardens can be extremely difficult to manage.

Plate 32. These chickens are cleaning up weed seeds in Steve Salt's garden prior to spring planting.

Plate 33. Snakes, such as this smooth earth snake (*Virginia valeriae*), can consume large numbers of slugs and soft-bodied herbivorous insects, as well as beneficial organisms such as earthworms.

Plate 34. Turtles, such as this female snapping turtle (*Chelydra serpentina*), are occasional visitors to temperate-zone gardens, where they may feed or lay eggs.

Plate 35. In wooded areas, tree frogs (*Hyla*) may be common, although they are more often heard than seen. This gray tree frog took up residence inside a hose reel on Steve Carroll's deck and was discovered when the hose was pulled out—giving the frog an unexpected ride!

Plate 36. Toads (*Bufo*) are workhorses of the garden, where they consume large numbers of insects and other invertebrates.

Plate 37. The Colorado potato beetle (*Leptinotarsa decemlineata*) is one of the more easily recognized beetle pests in gardens. Photograph by Scott Bauer; courtesy of the Agricultural Research Service, U.S. Department of Agriculture

Plate 38. Beetles do much of their damage as larvae (or grubs), when they feed on plant roots. Steve Carroll uncovered this grub while converting part of his lawn to a garden.

Plate 39. Caterpillars, although often strikingly beautiful, are major garden herbivores. Courtesy of George Shinn

Plate 40. Gardeners frequently plant nectar-rich flowers to attract adult butterflies, such as this giant swallowtail (*Papilio cresphontes*).

Plate 41. Butterfly larvae are herbivores. Unless the caterpillars find adequate food, however, they will not reach the adult stage appreciated by many gardeners.

Plate 42. Aphids are wingless during much of their life cycle, but winged stages, such as shown here, allow for migration. Photograph by Scott Bauer; courtesy of the Agricultural Research Service, U.S. Department of Agriculture

Plate 43. Scale insects not only damage plants directly, but they introduce secondary problems when molds, such as black sooty mold, grow on their exudates. Photograph by Scott Bauer; courtesy of the Agricultural Research Service, U.S. Department of Agriculture

Plate 44. Fruits are fed on by a wide variety of adult flies, including the notorious Mediterranean fruit fly (*Ceratitis capitata*). Photograph by Scott Bauer; courtesy of the Agricultural Research Service, U.S. Department of Agriculture

Plate 45. Bees are among the most important pollinators in both natural and cultivated landscapes. The honey bee (*Apis mellifera*) is widely imported and managed in an effort to bring crops to the dinner table and to market.

Plate 46. Spiders and their relatives are among the most common garden predators. Many familiar spiders, including the yellow garden spider (*Argiope aurantia*), trap prey in webs. Other arthropods, including daddy-longlegs (or harvestmen, family Phalangiidae) more actively search for food. Photograph of yellow garden spider courtesy of Michael Ira Kelrick

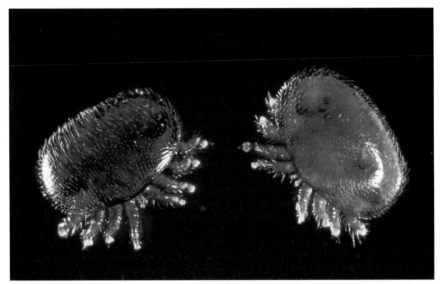

Plate 47. The varroa mite is a major cause of honey bee decline in North America and has resulted in considerable economic hardship among growers and beekeepers. Photograph by Scott Bauer; courtesy of the Agricultural Research Service, U.S. Department of Agriculture

Plate 48. Slugs cause extensive damage to herbaceous plants, especially to tender seedlings.

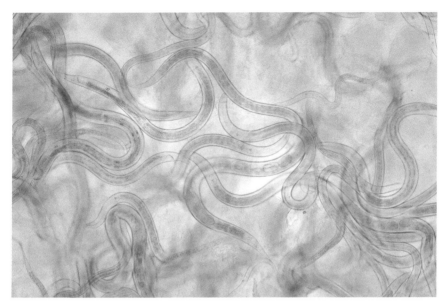

Plate 49. Nematodes exist in untold numbers in most soils. They play ecological roles ranging from herbivore to predator to parasite. Most nematodes are less than 1/25th inch (1 mm) long. Courtesy of Anne Bergey

Plate 50. Algae frequently colonize soils in gardens and greenhouse pots, especially soils that receive sufficient light but that do not dry out.

Plate 51. The scrambled egg slime mold (*Fuligo septica*) obtains nutrients by ingesting solid particles of organic matter. Courtesy of Anne Bergey

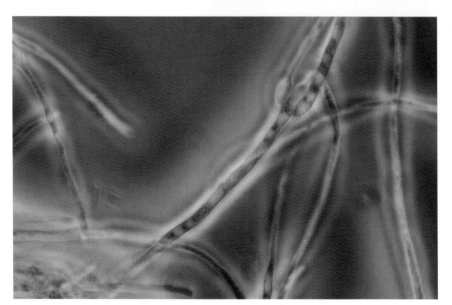

Plate 52. This photomicrograph shows the tubular hyphae of the fungus *Penicillium charlesii*.

Plate 53. Mushrooms such as this shaggy mane (*Coprinus comatus*) are the spore-bearing fruiting structures of fungi. Most of the mass of the fungus is below the soil.

Plate 54. Wood-rotting fungi help break down resistant molecules such as cellulose and lignin.

Plate 55. *Myxococcus xanthus*, a common Gram-negative soil bacterium, is shown here exhibiting characteristic swarming behavior (×4000). Courtesy of Brock Neil

Plate 56. Day length affects many aspects of plant growth and reproduction. This lettuce plant has bolted in response to increasingly longer days and shorter nights.

Plate 57. The mist hanging over this mountain meadow helps to keep plant tissues moist and moderates changes in air temperature.

Plate 58. Particulate matter, for example smoke from burning vegetation, can have a substantial effect on Earth's energy balance by blocking incoming solar radiation and reducing outgoing infrared radiation. Courtesy of Michael Ira Kelrick

Plate 59. Strong winds have damaged plants at the edge of this corn/maize field. Plants growing at the center of the field were protected from the wind and were undamaged.

Plate 60. This Vermont valley fog resulted from a combination of factors. Cold, dense air flowed downslope, and because cold air holds less water vapor than warm air, water droplets condensed and formed fog in the valley.

Plate 61. Low soil moisture, especially in combination with wind, can quickly wilt water-hungry plants such as squash.

Plate 62. Ice storms are a major source of limb loss and mortality in northern regions. Branches of this tulip tree (*Liriodendron tulipifera*) are completely encased in ice. Courtesy of Anne Bergey

Plate 63. Natural ponds are often complex environments, with floating, emergent, and shoreline vegetation supported by nutrients brought in through runoff, precipitation, and living organisms.

Plate 64. When clay soils dry out, they shrink, causing drought cracks that can damage and dehydrate plant roots.

Plate 65. This sheer road cut through loess soil along the Missouri River has resisted erosion due to its high silt content.

Plate 66. This profile through a rich, prairie soil has a dark, fertile A horizon that is up to 16 inches (40 cm) thick.

Plate 67. Seedlings require considerable energy to support their dramatic growth. They obtain this energy by the respiration of fats and carbohydrates stored in endosperm or cotyledons, shown here still attached.

Plate 68. Soil erosion can result in tremendous losses of energy and nutrients from gardens, as well as injury to plants.

Plate 69. These Cape buffalo (*Syncerus caffer*) are an example of one link in a food chain. They are shown here eating water hyacinth (*Eichhornia crassipes*), which has invaded waterways throughout much of the world. In turn, young or diseased Cape buffalo are preyed on by lions and other large predators, but scavengers, bacteria, and parasites commonly form the next link in this food chain. Courtesy of Anne Bergey

Plate 70. Materials leave gardens in many ways, through erosion, decomposition, herbivory—and through harvesting. Courtesy of Anne Bergey

Plate 71. Although not responsible for lawn care, Steve Carroll's dog has provided significant amounts of nitrogen and other nutrients. The grass has responded to this nutrient input with increased growth and color, but has also been burned in places.

Plate 72. When excessive nutrients—especially phosphorus—reach waterways, they often cause excess growth, or blooms, of algae.

Plate 73. Growth of plants is often limited by access to water. Providing adequate water to these plants would not lead to unlimited growth, however, because another require-ment, perhaps nitrogen, would eventually become limiting.

Plate 74. Western skunk cabbage (*Lysichiton americanus*) is unusual in that it generates its own heat, often while still under the winter snowpack. Courtesy of Pete Goldman

Plate 75. Temperate fruit trees, such as this sour cherry (*Prunus cerasus*), break bud in spring only after satisfying specific chilling requirements.

Plate 76. Plants growing side by side in a garden can vary dramatically in their tolerance of low temperature. A hard frost has killed the cold-sensitive luffa (*Luffa acutangula*) growing on the fence, whereas the more tolerant edible chrysanthemum (or shungiku, *Chrysanthemum coronarium*) has suffered only minor damage and the robust arugula (*Eruca sativa*) continues to thrive.

Plate 77. Gardeners rely greatly on the USDA Plant Hardiness Zone map in making plant selections; the map divides North America into 11 hardiness zones based on average annual minimum temperatures. Courtesy of the U.S. Department of Agriculture

Plate 78. If not protected, very tender tropical plants such as bananas suffer chilling injury at temperatures well above freezing. Courtesy of Anne Bergey

Plate 79. Even plant species normally able to tolerate quite low winter temperatures can be damaged if temperatures drop suddenly, as was the case with this peach tree.

Plate 80. These corn/maize seedlings were grown under three different light regimes. From left to right, they received high light, low light, and no light. The plant grown under low light is etiolated, that is, elongated as it stretches toward available light.

Plate 81. If not provided with adequate cover, peppers can sunburn quite quickly.

Plate 82. Saplings are prone to damage from wind, yet they benefit from moderate vibration. Therefore, it's important to stake young trees properly to allow some movement.

Plate 83. Plants vary widely in their abilities to tolerate drought. The sacred lotus (*Nelumbo nucifera*) is an aquatic plant with high water requirements, whereas cacti and other species of arid habitats have evolved mechanisms to survive extended periods of drought. Photograph of sacred lotus courtesy of Jeffrey M. Osborn

Plate 84. Even when thoroughly watered, transplants sometimes wilt as a result of damage done to their root systems during handling, as was likely true for this cinquefoil (*Potentilla*). Transplanting late in the day and providing shade can reduce transpiration and decrease the likelihood of wilting.

Plate 85. As a result of oxygen deprivation, too much water can kill or damage plants even more quickly than too little water. These sickly broccoli plants have been injured by waterlogging of their root zones resulting from flooding.

Plate 86. The highly allergenic wind-blown pollen of ragweed (*Ambrosia*) is the bane of hay-fever sufferers, but ragweed is among the world's most efficient plants at converting carbon dioxide into plant tissues.

Plate 87. The stunted, yellow-green corn/maize (*Zea mays*) plants in the foreground are showing symptoms of nitrogen deficiency. The tall, dark green plants behind them, which were planted at the same time, apparently have access to sufficient nitrogen.

Plate 88. Pumpkins are fast-growing plants that flower profusely and bear massive fruits with many large seeds. It may not be surprising that pumpkins require large amounts of phosphorus, as discussed in the text.

Plate 89. Lack of available calcium, especially under conditions of fluctuating moisture supply, can lead to severe breakdown of tomato fruit tissues, a condition known as blossom end-rot.

Plate 90. Rhododendrons are exceptionally tolerant of acidic soil and grow best at a pH below 5.5. Soil amendments such as flowers of sulfur are often added around rhododendrons, azaleas, and their relatives to increase soil acidity.

Plate 91. The leaves of this grapevine (*Vitis*), growing in an alkaline desert soil under irrigation, are showing symptoms of serious deficiencies of both iron and zinc.

Plate 92. This surf on Cape Cod, Massachusetts, is creating large amounts of salt aerosols (sea spray) that may be carried far inland by the wind, causing salt damage to susceptible plants.

Plate 93. Gardens often form diverse communities comprised of many kinds of organisms. This southwestern garden includes typical garden plants such as cosmos, as well as corn/maize, more often planted for food than as an ornamental. Even though planted along a city sidewalk, this garden attracts a variety of insects.

Plate 94. This landscape, on the campus of Truman State University in northeastern Missouri, includes considerable vertical layering and horizontal variation. These are achieved through plantings of bluebells (*Mertensia virginica*), shrubs of various statures, small *Magnolia* and redbud (*Cercis canadensis*) trees, and taller trees set in the matrix of a lawn.

Plate 95. Ecotones are transitions between adjacent communities, such as between this aspen (*Populus tremuloides*) forest and meadow. Ecotones often include unique species, as well as species found in each of the adjacent communities.

Plate 96. When jack pine (*Pinus banksiana*) forests burn, pine seeds germinate quickly in the exposed mineral seedbed, leading to re-establishment of the pine forest.

Plate 97. Through efforts of government agencies, bison increased in the greater Yellowstone ecosystem during the late twentieth century. More controversial has been a program to reintroduce wolves to this same ecosystem. Courtesy of Anne Bergey

Plate 98. In this garden, heat-sensitive strawberry plants benefited from the shade cast by taller okra plants. By the time the frost-sensitive okra died in the fall, the strawberries were well established. This is one example of companion planting discussed in chapter 6.

Plate 99. Gardens can be thought of as islands of selected plantings set in a matrix of pavement, lawn, or other plants. These formal beds at Roanoke Island Garden accentuate the island nature of gardens.

Plate 100. This view of Hawaiian taro (*Colocasia esculenta*) fields demonstrates the relationship between garden shape and perimeter, which has an important impact on dynamics at the garden edge. Courtesy of George Shinn

Plate 101. A mouse eating an acorn is an example of herbivory in action. However, while eating, the mouse must be alert for its own predators. Courtesy of Al Cornell

Plate 102. Yellow nutsedge (*Cyperus esculentus*) is a troublesome, allelopathic weed throughout much of North America, especially in poorly drained areas. Chufa is the edible tuber of a cultivated variety of this plant.

Plate 103. Field bindweed (*Convolvulus arvensis*) is a noxious weed that can quickly grow up and over garden plants.

Plate 104. The Old World climbing fern (*Lygodium microphyllum*), a quickly growing vine, has become a major problem in Florida. Photograph by Peggy Greb; courtesy of the Agricultural Research Service, U.S. Department of Agriculture

Plate 105. Although used as a potherb in Mexico and much of the Old World, purslane (*Portulaca oleracea*) is now more recognized in North America as a troublesome weed by virtue of its ability to tolerate drought.

Plate 106. Weeds are characteristically good at colonizing temporary habitats, which they accomplish through a variety of means. Shown here is butterfly weed (*Asclepias tuberosa*), whose seeds, like those of the more weedy milkweeds, are wind dispersed.

Plate 107. This pure stand of tulips, in Mount Vernon, Washington, is a dramatic example of a situation in which intraspecific competition might be expected to operate. Courtesy of Anne Bergey

Plate 108. Despite being one of the most aggressive weeds in the United States, purple loosestrife (*Lythrum salicaria*) is still sometimes planted for its strikingly beautiful flowers. Courtesy of George Shinn

Plate 109. Herbivory can cause a range of damage, from the chewing of leaves by this gypsy moth (*Lymantria dispar*) caterpillar to consumption of whole seeds by rodents. Photograph by Scott Bauer; courtesy of the Agricultural Research Service, U.S. Department of Agriculture

Plate 110. The top of this plant was removed by a feeding deer, and it is simultaneously being fed on by a spittlebug (family Cercopidae) from within a protective froth.

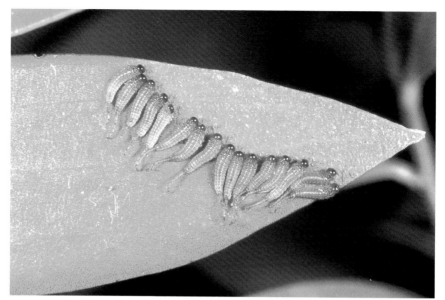

Plate 111. Sawfly larvae, shown feeding on a melaleuca leaf, often find younger, less well defended leaves more palatable than the older leaves on the same plant. Photograph by Jason Stanley; courtesy of the Agricultural Research Service, U.S. Department of Agriculture

Plate 112. Because grasses add new growth from below, horses remove older, tougher tissue as they feed, doing little injury to plants if not forced to overgraze.

Plate 113. Woody plants such as this peach tree, which was heavily browsed by sheep, are susceptible to injury or death if the vascular cambium, located below the bark, is damaged. This tree died a few months after the photograph was taken.

Plate 114. This pumpkin plant has been totally overrun by squash bug (*Anasa tristis*) nymphs, which will severely debilitate the plant, if not kill it outright.

Plate 115. The larvae of stem-boring beetles can quickly kill plants in which they feed. Often, the best treatment is to remove infected plant parts. Photograph by Ray Carruthers; courtesy of the Agricultural Research Service, U.S. Department of Agriculture

Plate 116. This sugar beet (*Beta vulgaris* var. *saccharifera*) has been extensively damaged by root-knot nematodes. Photograph by Scott Bauer; courtesy of the Agricultural Research Service, U.S. Department of Agriculture

Plate 117. Insects induce plants to form galls by stimulating overproduction of plant growth hormones or hormone mimics. This goldenrod gall has been pecked open by a bird looking for the insect larva inside.

Plate 118. Worldwide, weevils such as the maize weevil (*Sitophilus zeamaise*) cause extensive damage in stored grains. Photograph by Peggy Greb; courtesy of the Agricultural Research Service, U.S. Department of Agriculture

Plate 119. Bees consume large amounts of pollen and nectar, but also transfer pollen while they forage. Without this service, many flowering plants, such as this thistle (*Cirsium*) would be unable to reproduce successfully.

Plate 120. Like many plants, this sunflower (*Helianthus annuus*) is protected against some herbivores by stiff hairs.

Plate 121. Seeds represent the next generation and are often provided considerable protection. Shagbark hickory (*Carya ovata*) seeds are contained within the thick, tough walls of the hickory fruit.

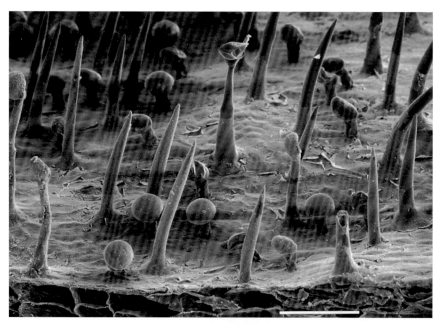

Plate 122. This scanning electron micrograph of a geranium stem shows two types of hairs. The long, sharp hairs provide physical defense, whereas the short, glandular hairs exude a chemical that repels herbivores (×200). Courtesy of Jeffrey M. Osborn

Plate 123. Monarch butterfly (*Danaus plexippus*) larvae take advantage of the defensive chemicals in milkweed (*Asclepias*) plants. The larvae sequester cardiac glycosides from the milkweeds within their tissues and use them in defense against their own predators. Courtesy of George Shinn

Plate 124. Ladybird beetles of various genera consume large numbers of aphids and are often introduced into gardens and greenhouses for this purpose. Photograph by Scott Bauer; courtesy of the Agricultural Research Service, U.S. Department of Agriculture

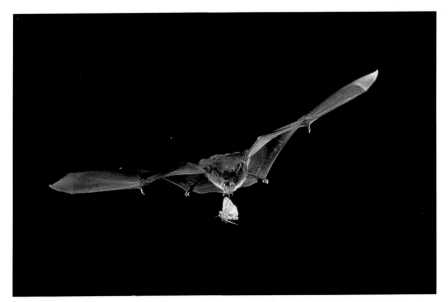

Plate 125. Bats, such as this big brown bat (*Eptesicus fuscus*), are common predators in and around gardens. They should be encouraged due to the large numbers of insects that they consume. Courtesy of Merlin D. Tuttle, Bat Conservation International

Plate 126. This green anole lizard (*Anolis carolinensis*) relies greatly on vision to detect its prey, which includes insects and spiders. Courtesy of Anne Bergey

Plate 127. Web-building spiders detect prey by sensing vibrations carried by the web's silken strands.

Plate 128. Many animals, such as this spider, take advantage of cryptic coloration to blend into the background, concealing them both from their prey and their predators. Courtesy of Michael Ira Kelrick

Plate 129. This northern walking stick (*Diapheromera femorata*) benefits from mimicry through its resemblance to the twigs of plants.

Plate 130. Mimicry of the monarch butterfly provides protection to this viceroy butterfly (*Limenitis archippus*) by virtue of the monarch's reputation for being distasteful.

Plate 131. Caterpillars that congregate in large numbers may benefit by decreasing their individual odds of being consumed. In some cases, numbers build up to the point where predators cannot consume all of the prey, ensuring that at least some will survive.

Plate 132. Zinnias and related plants may suffer from powdery mildew, a disease caused by the obligate parasitic fungus *Erysiphe cichoraceum*.

Plate 133. The spore-forming fruiting bodies of cedar-apple rust fungus (*Gymnosporangium juniperi-virginianae*) growing on junipers are deceptively attractive, appearing like orange ornaments.

Plate 134. Dense mats of fine hairs on the stems and leaves of plants, such as on these tomatoes, may pose formidable barriers to herbivorous insects, parasitic fungi, and invertebrate vectors of pathogenic microbes.

Plate 135. Some viruses are capable of transmission via seeds, as in the plum pox virus infecting this stone fruit. Photograph by John Hammond; courtesy of the Agricultural Research Service, U.S. Department of Agriculture

Plate 136. This wilting pepper plant is suffering from infection by the soil-dwelling *Verticillium* fungus, which has caused its vascular system to become plugged.

Plate 137. *Botrytis cinerea* gray rot causes major losses to strawberry growers. Photograph by Scott Bauer; courtesy of the Agricultural Research Service, U.S. Department of Agriculture

Plate 138. Hollow trees generally result from decay of the nonliving heartwood by wood-rotting basidiomycete fungi.

Plate 139. Severe, irreversible wilting of cucumbers or melons is often a symptom of the deadly bacterial wilt disease.

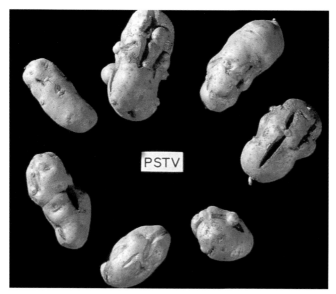

Plate 140. These potato tubers are misshapen as a result of
infection by the potato spindle tuber virus (PSTV). Photograph
by Barry Fitzgerald; courtesy of the Agricultural Research Service,
U.S. Department of Agriculture

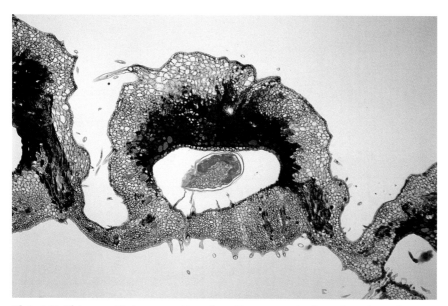

Plate 141. This magnified cross-section of a gall on a grape leaf reveals an insect larva
lurking inside. Courtesy of Jeffrey M. Osborn

Plate 142. This parasitoid wasp, a short-tailed ichneumon (*Ophion*), is laying its eggs in a gypsy moth (*Lymantria dispar*) caterpillar. Photograph by Scott Bauer; courtesy of the Agricultural Research Service, U.S. Department of Agriculture

Plate 143. This heavy infestation of dodder (*Cuscuta*) on jewelweed (*Impatiens capensis*) almost obscures its host.

Plate 144. The lumpy growths (or nodules) on the roots of these clover (*Trifolium*) plants are indicative of infection by mutualistic, nitrogen-fixing *Rhizobium* bacteria.

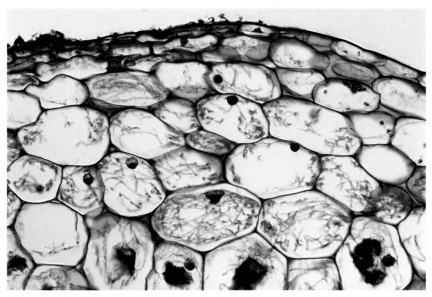

Plate 145. This light micrograph of an orchid root shows a mycorrhizal fungus invading host plant cells. Courtesy of Jeffrey M. Osborn

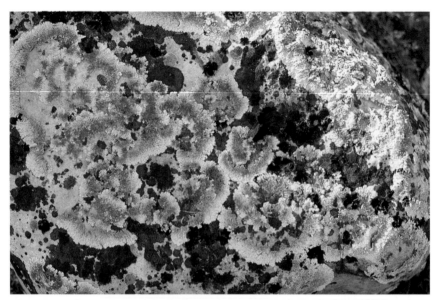

Plate 146. Lichens are primary colonizers of exposed or nutrient-poor locations.

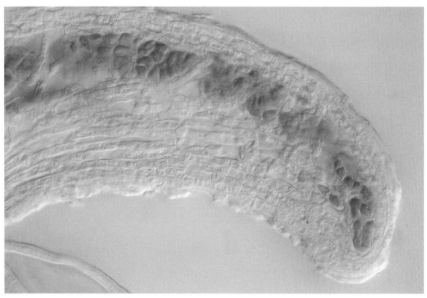

Plate 147. This micrograph shows a cross-section of a lichen with algal and fungal zones (×200). Courtesy of William B. Sanders

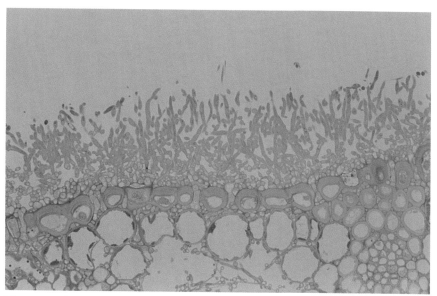

Plate 148. This micrograph shows reproductive structures of an endophytic fungus exposed in the tissues of its host, a grass plant. Courtesy of Jeffrey M. Osborn

Plate 149. Ants often tend aphids, much in the same fashion as humans tend domesticated livestock. In this mutualistic relationship, the aphids gain protection from predators and the ants consume the aphids' sugary exudate.

Plate 150. A goat browsing on tough, fibrous plants on this arid tropical hillside would soon starve without the aid of gut-dwelling organisms that convert the consumed plant tissues into nutrients that he can use.

Plate 151. Terracing gardens and lawns is one way to prevent soil erosion.

Plate 152. Careless irrigation, such as these sprinklers watering pavement, can waste large quantities of precious water.

Plate 153. Trickle irrigation is a highly efficient system of supplying water to garden plants.

Plate 154. During flooding of gardens by heavy rains, the elevation provided by raised beds combined with trenches may spell survival rather than death for plants.

Plate 155. Juniper berries serve as a food source for many animals, including cedar waxwings (*Bombycilla cedrorum*), yellow-rumped warblers (*Dendroica coronata*), and mice.

Plate 156. Many birds are irresistibly drawn to fruit crops and may devastate them unless foiled by barriers such as the netting draped over this cherry tree.

Plate 157. Sweeping a field or garden patch with a net is one technique for assessing the types and abundance of insects present. Courtesy of Gempler's, Inc.

Plate 158. Beekeeping is an ancient art that furnishes tremendous benefits, including honey and enhanced pollination of a wide range of garden plants.

Plate 159. Dill (*Anethum graveolens*) flowers serve as nectar sources for many beneficial parasitic and parasitoid wasps.

Plate 160. Sticky red balls may lure insect pests of fruit trees to a miry doom. Courtesy of Gempler's, Inc.

Plate 161. Traps laced with chemical pheromone mimics may exert a fatal attraction for pests. Courtesy of Gempler's, Inc.

Plate 162. This commercial strip is loaded with larvae of the parasitoid wasp *Encarsia formosa*, which will mature to attack greenhouse and nursery pests.

Plate 163. Floating row covers effectively exclude many insects from garden plants and help protect them from frost as well.

Plate 164. These tomato variety tags indicate genetic resistance to nematodes (N), *Verticillium* fungus (V), and *Fusarium* fungus (F).

Plate 165. Ground covers, plant covers, and supports such as cages can be combined to reduce infection of plants by pathogenic fungi and other disease organisms by preventing contact with damp soil and air-borne or rain-splashed spores and by reducing disease-promoting leaf dampness.

Plate 166. Seeds may be colored with bright dyes as a warning of fungicide or insecticide treatment.

Plate 167. Although somewhat less accurate than laboratory testing, economical soil test kits are available for use by gardeners and farmers. Courtesy of Gempler's, Inc.

Plate 168. Adding composted plant residues to gardens can be a highly effective technique for recycling nutrients and improving soil structure. Courtesy of the U.S. Department of Agriculture

Plate 169. This gardener is using composted steer manure (analysis 2–1–2) as fertilizer. Courtesy of the U.S. Department of Agriculture

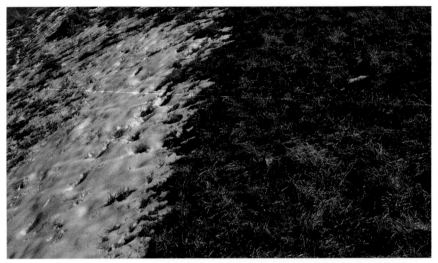

Plate 170. The angle of incoming sunlight can strongly influence its ability to warm the ground, as shown along this east-west ridge. Snow has melted on the south-facing but not on the north-facing slope.

Plate 171. This ornate tile-and-masonry garden bench absorbs substantial heat from the sun during the day and will gradually release heat at night, thus warming nearby plants.

Plate 172. Rotary tillers can be very effective and convenient garden tools, but they should be used minimally and kept well maintained to reduce their environmental impact. Excessive tillage, especially in the autumn, can reduce earthworm populations.

Plate 173. Dirty garden tools, such as these shovels, may serve as vectors for spreading disease organisms from one area to another.

Plate 174. Even gardens that really are islands, such as this one at the Missouri Botanical Garden in St. Louis, are not truly isolated. Gardens are intimately connected to the surrounding habitat through the atmosphere, soil, water, and the movement of organisms. Those who approach gardening from an ecological perspective are likely to grow healthy plants and to contribute to the well-being of the natural world as well.

Moderate wind can stimulate plants to produce stronger tissues, but high winds commonly shred or tear off leaves and can break limbs or even main stems or trunks of plants. Strong winds abrade and lacerate plants with suspended debris or sand and can carry injurious sea-salt aerosols far inland. These effects are especially problematic for taller plants, as average wind velocity increases dramatically with increasing distance above the ground. A heavy snowfall (or, worse, an ice storm) can snap even massive, mature trees, especially if the storm hits before trees have finished shedding their leaves for the season (plate 62). Some brittle species of shade trees such as box elder (*Acer negundo*) and silver maple (*Acer saccharinum*) are especially notorious for losing major limbs in storms and should not be planted near buildings or power lines. The wind also carries fungal and bacterial spores great distances, making it a major vector for transmission of plant diseases.

Wind can even shape the architecture of plants. Woody plants in chronically windy areas, such as exposed mountain slopes and along seashores, exhibit characteristic growth forms caused in large part by wind. Conifers show a flagging, whereas many other woody perennials grow prostrate among rocks and in exposed areas. In some regions, seasonal winds can strongly influence local gardening practices. The warm winter Chinook along the Rocky Mountain front ranges of Alberta and Montana may make local gardening climates one or two USDA hardiness zones warmer than more southerly locations on the open plains in Saskatchewan, Manitoba, and the Dakotas. The cold, dry Mistral and hot, dry Sirocco winds of the Mediterranean littoral or the Santa Ana winds of southern California can wither lush garden vegetation overnight. Nor'easters in New England and hurricanes along the Atlantic and Gulf Coasts of the United States can drive salt spray from the ocean far inland, affecting garden plants and soils.

Weather refers to the day-to-day antics of the atmosphere, and climate the average or long-term atmospheric conditions of a region. Weather and climate are the sum of all atmospheric properties and phenomena and are rivaled only by soil in their impact on gardening. Weather phenomena ranging from heat waves and droughts to fogs and tornadoes may stimulate vigorous plant growth or bring an abrupt halt to the lives of garden plants. Climate will dictate which perennial plants can survive without microclimate modification by the gardener.

A gentle rain of small droplets will soak the ground more thoroughly, with less runoff, soil erosion, or spattering of debris, than will a hard-driving rain with large droplets. In gentle rains less water makes it to the ground, however, because more water is lost to evaporation in midair when rain droplets are small than when they are large. Rainfall and snowfall are the major mechanisms for clearing suspended particulate matter from the atmosphere, as water droplets or ice crystals coalesce around particles until they can no longer be held aloft by air currents.

Snow is a good insulator. A layer of snow on the ground may actually warm the soil by conserving heat and reduce soil temperature fluctuations that injure overwintering plants. In areas where winters are cold but snow cover is rare or unreliable, gardeners are well advised to mulch overwintering plants with leaves or other protective material. However, sometimes a heavy snow cover will keep soil too warm, leading to growth of cold-tolerant pathogenic fungi that can injure overwintering plants. Particularly notorious is the snow mold *Marasmius oreades*, which parasitizes snow-covered grasses such as wheat and turfgrasses and keeps a lot of golf course greenskeepers busy in the spring resodding damaged areas. From a gardener's perspective, sleet, glaze ice, and hail are unmitigated evils, of course, and can severely damage or destroy plants in minutes by shredding leaves, breaking stems, bruising or puncturing fruits, and chilling plants. Heavy ice storms may create spectacular displays of crystalline forests glittering in the sun with diamond-like points of light—but they also lead to the roar of chainsaws cutting up broken trees for weeks afterward.

Fog and dew can supply plants in coastal desert areas, such as Baja California and northern Chile, with a sixth of an inch (several millimeters) of moisture per day, enough to sustain some drought-tolerant plants where rainfall may be virtually nonexistent. However, fog may reduce sunlight intensity to the point where only shade-tolerant plants can thrive, even in the open, and dew greatly aggravates the spread and development of many foliar diseases of plants.

Natural Waters

Plants do not live in laboratory beakers of pure water, but rather in oceans, streams, ponds, lakes, marshes, groundwater, and soil water. There are dissolved, suspended, and floating materials that make natural waters physically and chemically different from pure water (plate 63). Due to dissolved acids or bases, the pH of natural bodies of water ranges from 2 to 10 (seawater is usually about pH 7.5). Other dissolved substances in natural waters include salts, sugars, chlorine, alcohols, and pigments. Water having a high salt concentration is a particularly hostile growth medium for most plants, and as salt concentration increases, progressively fewer and fewer species can grow well. Bodies of water also have bulk physical properties such as turbulence, flow, thermal fluxes, evaporation, precipitation, and aerosol formation; these have strong effects on the well-being of plants, especially aquatic and wetland species.

Natural waters commonly carry large quantities of suspended materials, which may include clay, humus, silt, algae, bacteria, diatoms, and pollen. Debris and hydrocarbons tend to float on the surface. Thus, natural waters are often cloudy; colored blue, green, yellow, brown, or orange; and generally more viscous and denser than pure water. These factors often affect the suitability of water for irrigation and may impact plant growth when flooding or runoff enters a garden.

As they pass through the atmosphere en route to Earth, rain and snow dissolve substances such as carbon dioxide and oxides of sulfur and nitrogen. Consequently, even "pure" rainfall is actually weakly acidic (about pH 5.5), and when polluted it may become even more strongly so. In an extreme case, a pH of 1.5 was recorded for acid rain high in hydrated nitrogen oxides. Thus, in all humid regions, acidification of soil will occur over time unless there are large reserves of limestone or other alkaline materials in the soil—or unless a gardener adds lime from time to time. In arid regions, where evaporation often exceeds rainfall, salts will accumulate so that soils generally remain neutral or alkaline and may even require acidification for good garden plant growth.

Atmospheric precipitation is ultimately dependent on evaporation from surface waters, soil, and plants, and any significant change in evaporation rates will influence precipitation. Especially in humid, tropical areas and during the summer in warm, humid, temperate regions, a large percentage of total rainfall results from local recycling of moisture rather than from moisture evaporated from the oceans. One important consequence of this phenomenon can be seen when local rainfall declines dramatically upon clear-cutting of tropical rainforests or drainage of wetlands. This drought effect may also cause irreversible dehydration of high-clay soils and render the area a permanent desert. Conversely, planting vegetation in a barren area may actually increase local precipitation.

Soil

People walk on soil every day, but few of us give it much thought. Yet, virtually all life ultimately depends on this thin skin of the Earth. Terrestrial plants depend on the soil for mechanical support, as a thermal buffer and mechanical protector, as a habitat for essential symbiotic organisms, and as a source of water and nutrients. Even in the oceans, most mineral nutrients ultimately come in runoff from land masses.

But just what *is* soil? And what properties does it have that make it what it is? Soil is not simply powdered rock. No matter how fine, powdered rock does not have the characteristic properties of soil. Fragments of native rock are often found in soil, but the essential nonliving components are a variety of clay minerals, humus, and small molecules and ions produced by microbial and plant metabolism. Clay minerals are complex, sandwich-like assemblages of materials that are derived from rock minerals and the atmosphere by the combined actions of microorganisms, water, acid, and weather. Humus, a substance formed only in soil, is a stable complex of electrically charged, very large organic molecules ultimately derived from the remains of dead plants, animals, and microbes. Humus and, to a lesser extent, clay bear high densities of electrical charges that are critical for adsorbing and storing plant nutrients and water. Other small molecules derived from soil-dwelling plants and microbes include nutrients, hormones, and toxins.

Soils may be distinguished by their colors, texture, and stickiness. Gardeners speak of "sandy soil," "red clay," "black loam," and "gumbo," for instance. (Loam is soil with a more-or-less balanced blend of sand, silt, and clay. Gumbo soil, common in parts of the American Midwest and South, has a very high content of certain clays that result in extreme stickiness and swelling when wet, then shrinking, cracking, and severe hardness upon drying.) Soils also have various characteristic odors and, yes, flavors—although few modern gardeners imitate old-time farmers who routinely tasted soil to evaluate its quality. Color, odor, and taste relate to a soil's chemical properties. Deep brown or nearly black soils are generally high in humus and clay and are fairly fertile. Unfortunately they also tend to be prone to shrinking and swelling, are susceptible to compaction, and often have hardened clay or lime layers. Sandy, pale soils from conifer forests are usually quite acidic and infertile. Red or orange clay soils of the tropics and subtropics tend to be moderately acidic, phosphorus-deficient, and rather infertile, but they do not shrink or swell and are surprisingly nonsticky when wet. Pale gray or whitish soils in desert regions are often quite alkaline and/or salty. Blue, bluish gray, or yellowish gray mottling against a yellowish background, especially with a sour odor, indicates periodic waterlogging of a soil. A strong earthy aroma, on the other hand, is indicative of abundant actinomycetes, a bacterial group most numerous in fertile, well-drained soils.

Many of a soil's distinctive properties that enable it to support plant and microbial life and neutralize toxic substances are conferred by clay and humus. Humus is particularly capable of absorbing and releasing water and water-soluble plant nutrients, as well as water-repellent substances such as oils and pesticides. Clay minerals are less adept at adsorbing water-repellent substances, but can reversibly bind considerable amounts of water and ions. In chemical terms, soils, especially those rich in clay or humus, are extremely capable buffers for a wide range of substances, stabilizing the growing environment of plants against wide or abrupt changes in chemical or physical properties. This powerful buffering capacity of soil is both good and bad news for gardeners. Soils can act as high-capacity reservoirs of acid or alkali, nutrients, and water. They can also store tremendous quantities of undesirable substances. Consequently, it is usually a slow, difficult process to fundamentally alter most chemical properties of soil. For example, a soil that is too acidic may require repeated applications of many tons per acre (hectare) of lime to even slightly and temporarily make it more alkaline.

Another significant property of soils rich in certain clay minerals is that they dramatically shrink when they dry and swell when they are wetted. This shrinking and swelling can disrupt root systems of plants, aggravate dehydration of the soil during drought, worsen runoff during heavy rains, and cause buckling of roadbeds and building foundations (plate 64). Many clay-rich soils also undergo frost-heaving, when columns of soil expand upward as they freeze, then subside as they thaw. This heaving can actually lift shallow-rooted plants right out of the ground, which is a good reason for mulching perennials in the winter to maintain a more constant soil temperature.

Under anaerobic conditions, such as those resulting from waterlogging or compaction, soil chemical properties may change dramatically from those of well-aerated soils. Elements such as iron, manganese, and copper may be converted to more reduced (less positively charged) forms that are generally more soluble than the oxidized (more positively charged) forms found in aerated soils. These reduced metals are the source of the characteristic green, blue, gray, and pallid yellow colors observed in anaerobic soils. Anaerobic conditions also dramatically alter populations of microbes in the soil, slow the decomposition of organic matter, and lead to the production of organic acids, alcohols, and noxious gases such as hydrogen sulfide, methane, volatile amines (including aptly named putrescine and cadaverine), carbon monoxide, and hydrogen cyanide.

Soil also possesses other physical properties of which gardeners should take note. For example, wetting and drying of soil are not exactly reversible processes. Due to the physical chemistry of clay and humus, it's far easier to further wet soil that is already damp than it is to dampen completely dry soil. This is one reason why potting mixes should be moistened before being put into containers, and also why trickle irrigation, which supplies a slow, continuous flow of water to plants is more efficient than periodic flooding from hoses or furrows. In fact, if some tropical soils are dehydrated strongly enough (as by baking in the sun) they may become very water-repellent, a fact of woeful significance after deforestation of tropical rainforests.

Soil contains mineral particles ranging in size from submicroscopic clay crystals to boulders bigger than houses. The relative proportion in soil of these various particles (soil texture) has profound implications for its physical and chemical properties. Consequently, the U.S. Department of Agriculture, strictly using the metric system, has developed a standardized system for classifying soils based on percentages by weight of clay (particles less than 0.002 mm in diameter), silt (0.002–0.05 mm), sand (0.05–2 mm), and stones (greater than 2 mm). Classes such as silty clay loam or sandy clay may seem somewhat arbitrary, but actually correspond fairly well with performance of soils under agricultural or horticultural use. Soils high in clay tend to hold large amounts of water and nutrients, but they are difficult to work either wet or dry, warm up slowly in the spring, are susceptible to waterlogging, dry out slowly after rain or irrigation, and are quite susceptible to compaction. Turfgrasses and corn/maize generally do

well in soils high in clay. Sandy soils have somewhat the opposite properties and are suited to drought-tolerant plants or root crops such as carrots
grown under irrigation. Soils high in silt are rarer; their overall properties
are intermediate between sands and clays, although they often have the
unique characteristic of maintaining their shape when cut or dug. Some
sheer road cuts through high-silt loess along the Missouri River have stood
for decades without significant slumping or erosion (plate 65). Loam, with
its balanced blend of particle sizes, is generally the most favorable soil type
for raising garden plants.

A crude but simple technique for estimating soil texture that is accurate enough for garden usage is to roll a ball of damp soil about 2 inches (5
cm) in diameter in the palm of your hand and then gently crush the ball. If
the ball easily crumbles into more-or-less separate pieces, it is high in sand.
If the ball only flattens and deforms, the soil is high in clay. If the ball cracks
and splits as it flattens but does not fall apart, it is high in silt or is a loam
soil. Soil scientists, farmers, and experienced gardeners often become quite
adept at assessing soils using the ball method.

Sand, silt, and clay particles are generally not found in free form in
soils, but are organized into larger, cohesive masses called aggregates, especially in soils high in clay or humus. Aggregation of soil has profound consequences for the performance of soil as a plant growth medium. Most gardeners have noticed that soil tends to stick together in chunks of various
sizes and shapes (clods). What is less often noticed is that clods tend to
have fairly distinctive shapes and dimensions and that near-microscopic
aggregates bind together to form larger aggregates, which in turn coalesce
into even larger structures, and so forth. This aggregation of soil into discrete lumps is what fluffs up soil and prevents it from packing under its
own weight into a dense, anaerobic, virtually impenetrable mass. This
becomes painfully apparent when soil aggregation is destroyed by sodium
salts or by physical compaction. Adobe, pottery, and bricks are generally
made by strongly dehydrating high-clay soil that has had its aggregate
structure deliberately destroyed by kneading, compaction, or treatment
with salt water.

If a gardener digs a deep hole and observes its walls, the three-dimensional vertical structure of soil is often apparent. The top layer (horizon) of
many soils is a relatively thin layer of partially decomposed plant materials
called the O horizon (for "organic"). Next there is usually an A horizon of

mineral soil mixed with plant roots, decaying vegetation, and humus; this horizon varies from a few inches to many feet in thickness (plate 66). In temperate areas of moderate rainfall, the A horizon may be quite dark brown and popularly indicative of rich soil. Together, the O and A horizons form the topsoil. Below the A horizon is the B horizon, sometimes called the subsoil. The B horizon contains deposits of materials such as clay and lime that have been washed out of the A horizon, as well as plant roots (although many fewer than in the topsoil). Often the subsoil will be relatively infertile, dense, high in clay or rock fragments, and difficult to work. Below the B horizon is the C horizon, which may be bedrock, glacial till, wind-blown loess, sand, or other stuff from which the biologically modified true soil above (horizons O through B) has been derived. Few gardeners, except perhaps those in mountainous or hilly areas of very thin soils, will encounter the C horizon, although it often plays an important part in determining many properties of the soil above it.

Unfortunately, due to aggravated soil erosion or disturbance, many gardeners discover that the subsoil is their only available soil. Backyard gardeners in new subdivisions are especially unlikely to find intact topsoil, which is often removed or severely disturbed during house construction. Instead, gardeners in such locations are all too likely to discover a motley collection of sand, gravel, subsoil, nails, chunks of concrete, and miscellaneous junk, which some homeowners have whimsically named the D horizon—for "debris" or "dump."

The boundary between the A and B horizons may be surprisingly distinct, with a sudden change in color, texture, and other properties. In fact, this boundary is often demarcated by a dense layer of hard clay or precipitated lime (hardpan) or a lens of sand or gravel. Hardpans and lenses may pose formidable barriers to penetration of water, air, and roots into the subsoil below. After rain or irrigation, water may accumulate on top of a lens or hardpan, forming a perched water table that may waterlog the soil in surprising locations, such as on steep hillsides. During dry periods, plants may suffer from drought injury and nutrient deficiencies, even when the soil overall has adequate water and nutrients, because roots cannot penetrate through the dense barrier to the subsoil beneath.

Another result of the three-dimensional structure of soil is that energy (heat) and substances such as moisture and gases can move up, down, or sideways. This movement has interesting consequences. One is that deep soil layers may actually be warmer in winter than are surface layers, due to storage of heat from the previous summer combined with chilling of the surface layers. Conversely, deep soil layers often are cooler than the surface in summer. Another result is that water can move laterally if it encounters an obstacle during downward flow and even upward if it is wicked by evaporative dehydration of surface layers. Also, gases can move with surprising speed through sandy or well-aggregated soil, a fact of great significance for oxygen-requiring roots, as respiration by roots may locally deplete oxygen in the soil air, requiring its replenishment from elsewhere.

Energy in the Garden

When gardeners run gasoline-powered tillers through their gardens, apply fertilizer, or hoe up weeds, they're not only handling materials but also making a considerable input of energy. This may be obvious in the case of the tiller, but how so in fertilizing and hoeing? As considered in our discussion of the nitrogen cycle (chapter 4), the process of fixing atmospheric nitrogen into chemical forms available for plants demands much energy. In fact, an estimated 80 percent of the energy input into North American crop production is incurred, not in operation of fuel-gulping tractors and other power equipment, but in production of synthetic nitrogen fertilizers. Even phosphorus fertilizers, which undergo little change from natural mineral forms, require considerable energy for their mining, purification, and shipping. And the hoeing? Well, the swinging of the hoe is powered by the burning of blood sugar or reserve energy supplies (fat), which in turn originated by consumption of food that ultimately came from sunlight-driven photosynthesis.

The primary inputs of energy into garden ecosystems come in the forms of light, stored chemical energy, and heat. Other, generally much lesser, inputs include electricity and mechanical energy. Mechanical energy is input into the garden through such physical processes as the churning of soil by tiller tines, which not only rearranges plant and soil materials but also generates heat via friction. Chemical energy is stored in bonds between atoms; it can be released when those bonds are broken. For instance, the intense light and heat released by a roaring bonfire originate from the breaking of chemical bonds between atoms in hydrocarbons, cellulose, and other substances contained in the wood or other materials being burned. Let's take a closer look at what happens with cellulose.

During the burning, some of the energy released by breaking up cellulose molecules is consumed in reforming bonds between carbon and oxygen to produce carbon dioxide, and more is used for initiating the breakdown of yet more cellulose molecules—but overall there is a tremendous net release of uncontrolled energy, hence the concern of the fire department. Much of the chemical energy available in cellulose molecules originated from light energy used by plants' photosynthetic machinery to reform low-energy carbon dioxide and water into high-energy cellulose. When plants respire the sugars formed during photosynthesis, the chemical reaction that takes place is essentially the same as during the burning of wood. However, by using enzymes rather than a match to trigger the reaction, plants and other organisms are able to slow the burning (cellular respiration) to the point where much of the energy released can be used to drive energy-requiring processes such as synthesis of proteins, the fixation of nitrogen, and mechanical work (for instance, pushing root tips through the soil; plate 67).

Whatever the source or form of energy in a garden ecosystem, this energy rarely remains in its original form for long but swiftly changes in form, magnitude, and/or location. Changes in form occur via processes such as chemical reactions, friction, and phase changes (for example, when water freezes or evaporates). Some of the light energy absorbed by plants is changed to chemical energy via photosynthesis. The efficiency of this conversion is very low—plants are only able to store about 2 percent of the energy in sunlight as chemical energy in carbohydrates. Plant growth is limited in most cases, however, not by light intensity, but by heat, cold, disease, water deficiency, or shortages of mineral nutrients.

Heat can drive a wide range of chemical processes, as readers who have spent time with a Bunsen burner in a chemistry lab can testify. Plant growth and development rates are strongly tied to inputs of heat from the environment. Addition or removal of heat also drives physical processes such as evaporation, condensation, freezing of water, and thawing of ice.

Stored chemical energy (for instance, within the chemical bonds of carbohydrate or fat molecules) can be released to drive other chemical reactions, generate mechanical power and motion (such as movement of insects), or create light and heat. Energy stored in chemicals such as sulfates or nitrates is also harvested for life processes by anaerobic bacteria as they convert these molecules to water plus solid sulfur or gaseous molecular nitrogen, respectively.

Because energy can neither be created nor destroyed, changes in total energy content of a garden ecosystem can only occur by importing energy from or losing it to the universe outside of the garden—such as by sunlight shining on plants or water molecules evaporating and being blown away (along with their energy contents). Within a garden, energy essentially moves about freely. For example, substances or organisms moving from one place to another are accompanied by their stored chemical energy contents. Energy can also move via conduction, radiation, convection, electrical currents, and other means. Some of these movements may have surprising consequences for garden plants. For example, summertime heating of the soil surface results in a wave of heat energy that slowly moves downward through the soil by conduction. Similarly, loss of heat from the soil surface in winter results in a slow upward movement of heat from deeper levels. The relatively rapid heating and cooling of the upper soil, combined with the slow movement of heat to and from the deeper soil levels, often results in the deep soil being warmer in winter than in summer—an effect of great importance for deep-rooted perennial plants, whose roots may not go dormant in winter despite the near-dead appearance of the aboveground plants.

Because water can absorb extraordinarily large amounts of heat, when water moves, so does a considerable amount of heat. Thus, cold irrigation water may chill soils in summer more powerfully than does cold air in winter. On the other hand, warm water can heat soil and plants far more efficiently than can warm air—a fact used to advantage by some greenhouse growers during cold weather.

However it enters and whatever transformations it undergoes, all energy is eventually lost from physical systems, including gardens. The energy losses from garden ecosystems include exportation of heat and stored chemical energy. Heat radiates into the cosmos from plants and from the soil surface, as becomes all too apparent on clear, frosty autumn and spring nights. Heat also moves by conduction both downward and laterally out of the garden into the great sink of the Earth's mass, and heat accompanies water evaporating or flowing out of a garden. Farms and gardens may suffer greater loss of stored chemical energy than natural ecosystems because of the harvesting of plants and the often much greater rates of soil erosion (plate 68). Thus, the energy content of domesticated ecosystems must be continually replenished, most often through the addition of energy-rich nutrients.

Food Chains and Webs

When an organism consumes part or all of another organism, one link in a food chain is formed (plate 69). When an amoeba eats bacteria that decomposed a coyote that ate a rabbit that ate clover, for instance, or when a hawk preys on a robin that consumed a white cabbage moth caterpillar that ate a broccoli plant, these species form simple food chains. However, the hawk might also eat other birds and rodents; the robin might feed on invertebrates other than white cabbage moth caterpillars; the rabbit's sibling might be eaten by a snake that is eaten by the same hawk that ate the robin; and an omnivore such as a raccoon might eat a plant, a caterpillar on that plant, *and* a songbird or snake if it can catch one. Such links between food chains give rise to complex food webs, which tie together the many transfers of materials and energy in an ecosystem.

Consumption of one organism by another may take several forms—for example, herbivory, parasitism, predation, or decomposition of dead tissues—but the end result is that a consuming organism incorporates part of the ingested matter into its own tissues and breaks down part to generate energy for its activities and life processes. However, most energy and nutrients contained in ingested food end up as heat or waste products. Biologists estimate that, under ideal conditions, the average efficiency of conversion of ingested food into useful energy and living tissues is about 10 percent for herbivores eating plants, 25 percent for carnivores eating other animals (animal tissues are more concentrated and easily digested nutrient sources than are plant tissues), and 0.5–10 percent for decomposition of dead tissues by microbes. However, exact efficiencies vary widely with the particular organisms and environmental conditions involved, and they are usually much lower than ideal.

Because of this inefficiency, each level of a food chain is able to support a smaller amount of living tissue than can the previous level. Thus, 1 acre (0.4 ha) with 20 tons (18,000 kg) of plant tissues, if completely harvested and consumed under ideal conditions, could theoretically support 2 tons (1800 kg) of herbivores (about 4 cows or 120,000 caterpillars). These in turn could support 1000 pounds (450 kg) of carnivores (about 10 wolves or 20,000 wasps), which on their death could support 20 pounds (9 kg) of decomposers such as bacteria and fungi, and so forth. However, actual efficiency is generally far less than this, and a very high real-world yield of cows might be one cow per acre (2.5 cows per hectare) per year, so actual efficiency of conversion of biomass in a food chain is usually much lower. This inefficiency helps explain why food chains rarely extend past four levels. It also explains why plants are usually the limiting factors in biological productivity of any area—the capability of a patch of ground or water to support living organisms ultimately is proportional to the amount of plant tissue produced per season. Is it any wonder, then, that there are usually more insects and other organisms in a fertilized, well-watered, heavily planted garden spot than in a less well-endowed wild spot nearby?

Material Cycles

In any given garden, concentrations and forms of nutrients and other materials vary greatly in both space and time. Some substances undergo changes that are effectively irreversible, such as the slow transformation of rocks into clays and salts. Other changes occur more rapidly or may be reversible. Of particular importance in gardens are the reversible cycles involving water, carbon, nitrogen, and other plant nutrients—especially because activities of gardeners may strongly impact these cycles.

A material cycle is basically a series of transformations in which a substance begins and ends in the same form. There can also be cycles within cycles, in which intermediaries of the larger cycle undergo smaller cycles of their own. Material cycles in open ecosystems like gardens involve external sources of materials (for example, atmospheric carbon dioxide), internal reservoirs into which materials enter and are more or less permanently withheld from the cycle (plant tissues), and, usually, losses of materials that may or may not be reversible (a migrating herbivore).

Transformation of materials is the basis of chemistry. Therefore, to understand material cycles in garden ecosystems, an understanding of a few basic chemical terms is helpful. Electrons are negatively charged particles and protons are positively charged particles in atoms. When one or more electrons are lost from a substance, it then becomes more positively charged and is said to be oxidized. A substance that receives extra electrons from another substance becomes more negatively charged and is said to be reduced. Ions are atoms that have extra or missing electrons, so that they bear negative or positive electrical charges, respectively. Acids are substances capable of releasing free protons, whereas bases can absorb and bind free protons. When acids and bases come in contact with one another, they may react and neutralize one another to form salts and water.

Oxygen has a strong tendency to seize electrons from almost any other substance. Thus, under aerobic conditions, many elements (including carbon, phosphorus, sulfur, and most metals) tend to exist in oxidized forms, having lost one or more electrons to oxygen. Under anaerobic conditions, when free oxygen is scarce or absent, such as in stagnant water or in flooded or severely compacted soils, many substances exist in more reduced (electron-rich) forms, which have markedly different properties than their more oxidized versions. For example, oxidized iron compounds in soil are usually reddish brown or orange, odorless, and insoluble in water, whereas reduced ones are commonly green or yellow, have a metallic odor, and are much more soluble in water.

Chelating agents are organic (carbon-containing) molecules with clawlike cavities that tend to reversibly bind and stabilize charged molecules such as metal ions forming chelates. Water-soluble chelating agents may bind to substances, such as some metal ions, that otherwise would be insoluble in water, and thus are important in gardens for providing otherwise water-insoluble nutrients to plants.

The Hydrological Cycle

The global cycling of water mainly involves physical rather than chemical changes (that is, freezing and evaporating rather than breaking the bonds in the water molecules). In the limited confines of a single garden, however, there is very little local cycling of water. Instead, a garden is mostly a way station in the larger global cycle, although a very important way station. The primary inputs of water into most gardens are natural precipitation and irrigation. Other inputs include atmospheric humidity, groundwater, runoff, flooding, moisture present in compost and other imported materials, and water used to dissolve and apply fertilizers and pesticides.

Once in the garden, water is primarily stored in the soil, although a significant amount may be incorporated into plant and microbial tissues. Water that enters a garden is eventually lost, however; it may evaporate from the soil, run off, seep downward into the groundwater, or be removed when plants are harvested or weeded—although the greatest loss of water under most conditions is through transpiration from leaves and stems of plants. The dramatic effect of transpiration can be readily observed by comparing the relatively rapid speed with which soil dries out in pots that contain living plants as compared with soil in pots lacking plants.

Large-scale human activities, such as replacing vegetation with pavement in urban areas, draining or creating wetlands, or revegetating a large region, can have pronounced effects on the timing, amount, and intensity of precipitation. Individual gardeners can do little to influence the global hydrological cycle, but a gardener's activities *can* strongly influence the hydrological cycle on a smaller scale. For example, leaving ground bare of vegetation for an extended period (leaving ground fallow) may dehydrate the surface layer of soil and reduce atmospheric humidity, but it can also increase subsoil water content because of the absence of roots to remove water. Other gardening practices that affect moisture include choosing and spacing plants; enhancing or depleting soil organic matter; aerating or compacting soil, irrigating, and mulching.

The Carbon-Oxygen Cycle

All natural material cycles connect to some degree with one another, but those of carbon and oxygen interconnect so intimately that they may best be considered as subcycles of a larger combined cycle. The primary input of carbon into most gardens is carbon dioxide from the atmosphere. Other inputs of carbon may include carbon dioxide dissolved in rain, snow, and irrigation water; lime (calcium carbonate) used for neutralizing acidic soil; and organic matter such as compost, mulch, animal manure, and potting mixes. Oxygen is derived from atmospheric oxygen gas, carbon dioxide, and, via photosynthesis, from water. Applied materials containing cellulose, other carbohydrates, lignin from woody plants, and most other biological molecules also contribute appreciable amounts of oxygen, as do some inorganic chemicals (for example, nitrates and phosphates) and many synthetic substances.

Water contributes to the carbon-oxygen cycle when plants and microbes split water molecules into oxygen plus hydrogen during photosynthesis. The hydrogen thus obtained is used to convert carbon dioxide into sugars, and the oxygen is released as a by-product. Almost all free oxygen on Earth derives from this photosynthetic splitting of water. To complete the cycle, carbon-containing molecules, which derive from carbohydrates formed during photosynthesis, are rejoined with oxygen during respiration or combustion to generate carbon dioxide, water, and energy.

Carbon-containing molecules may be recycled many times through living organisms as they are ingested and incorporated into tissues or used as energy sources, being released again when organisms are eaten or decompose. With each turn of the cycle, however, part of the carbon returns to the environment as carbon dioxide. Were it not for continuous regeneration of organic molecules by photosynthesis, the whole cycle would eventually run down. In the absence of oxygen, as in compacted or water-saturated soils, a different carbon subcycle comes into play in which methane, carbon monoxide, and other reduced (electron-rich) carbon compounds are generated from biological molecules and then used as energy sources.

Another fate of carbon in garden ecosystems is to become humus—a relatively stable, complex, organic substance formed by microbes from the remains of plants and animals. It has a lifetime in the soil of decades, centuries, or even millennia, as contrasted with proteins and carbohydrates, which may be broken down by soil microbes in weeks, days, or even hours. Although only a small portion of dead organisms' tissues may be converted into humus, the durability of humus causes it to accumulate in soils to the extent that approximately half of all carbon in terrestrial ecosystems and the atmosphere combined exists as stabilized soil organic matter. (Much more carbon is dissolved in seawater or occurs in deposits such as coal, petroleum, and limestone.) Humus is extremely important in soil as a reservoir of slowly available nutrients and as an adsorbent of water, nutrients, and water-insoluble materials, including oils and many pesticides.

Human activities can greatly accelerate net loss of carbon from a garden ecosystem. These activities may include aggravating erosion, which exports humus-rich topsoil; tilling, which mixes oxygen into the soil and thus accelerates decomposition of organic matter; lighting fires, which can rapidly convert organic matter into smoke and carbon dioxide; and, probably most significant of all, wholesale removal of carbon as grazed, harvested, pruned, or weeded plant tissues (plate 70).

The Nitrogen Cycle

Of the nitrogen in the environment of plants and other land-dwelling organisms, approximately 55–60 percent is found in the atmosphere, whereas an additional 30–35 percent is chemically bound in humus. Most of the balance is in protein, DNA, and other molecules in living organisms. Only about 1 percent, mostly in the form of nitrates or ammonia and ammonium in soils and waters, is available for use by plants.

The most significant source of available nitrogen for plants in natural ecosystems is biological fixation of atmospheric nitrogen into nitrates and ammonia by symbiotic microbes such as *Rhizobium* and by free-living microorganisms, which we discuss in chapter 5. However, in modern times, synthetic fertilizers containing ammonia, ammonium salts, or urea—all generated from atmospheric nitrogen in factories powered by fossil fuels—have come to rival naturally fixed nitrogen substances, especially in intensively managed garden and farm soils. Regardless of how it happens, nitrogen fixation is an extremely energy-demanding process. This fact ties the nitrogen cycle to the carbon-oxygen cycle, as photosynthesis is the ultimate source of almost all energy used for nitrogen fixation. This is true whether the energy for nitrogen fixation is obtained through microbial respiration of plant carbohydrates or through industrial combustion of fossil fuels that were originally formed via photosynthesis eons ago. Other, lesser inputs of nitrogen into modern gardens include natural materials containing nitrate salts (for example, saltpeter or guano); manure or urine containing urea or uric acid; remains of plants and animals such as fishmeal and cottonseed meal; and nitrogen oxides from lightning discharges or acid rain.

Within gardens, nitrogen compounds undergo a wide range of cyclic changes and move between organisms, often quite rapidly. During fixation by microbes or factories, gaseous nitrogen is converted to ammonia. Ammonia can be absorbed by clay or humus, oxidized to nitrite or nitrate, or taken up and incorporated by plants or microbes into proteins, DNA, chlorophyll, and a host of other organic molecules. When plants die or shed tissues, their nitrogen-containing molecules can be reincorporated into other living organisms, converted into humus, or converted by microbes back to ammonia plus various carbon-containing substances. Animals excrete excess dietary nitrogenous materials as urea or uric acid. When added to soils from fertilizers or animal manure and urine, urea or uric acid is rapidly converted by microbes into carbon dioxide plus ammonia, which can then be recycled (plate 71).

Nitrates in soils can be taken up and incorporated by plants or microbes, reduced back to nitrite or ammonia, or leached into groundwater.

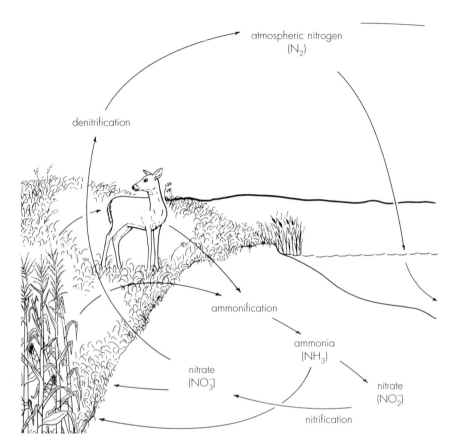

Figure 8. The nitrogen cycle represents one of the most complex and important nutrient cycles for gardeners (see the text for details). Inadequate nitrogen is often responsible for limiting plant growth in gardens, on farms, and in natural ecosystems.

In some rural and suburban areas that depend on groundwater-fed wells for drinking water, leaching of nitrates into groundwater has become a serious public health problem because free nitrates are toxic, especially to young children and pregnant women (causing so-called blue baby syndrome). However, the most significant ecological fate of nitrates is their large-scale reduction by microbes to nitrogen gas, which then returns to the atmosphere—thus completing the cycle started by fixation. Microbial conversion of nitrates to nitrogen gas is the primary cause of nitrogen loss from both garden and wild ecosystems and the main reason why available

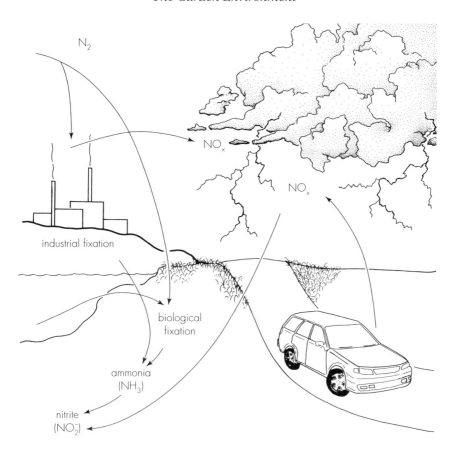

forms of nitrogen must continually be added to ecosystems for plant life to flourish. Nitrogen is also lost from gardens through evaporation of ammonia from alkaline soils, by removal of plant material by humans and animals, and via erosion of topsoil.

There are strong seasonal effects on the operation of the nitrogen cycle in gardens. In cold soils, conversion of organic forms of nitrogen, such as proteins, to soluble nitrogen salts is greatly reduced, and plants also have difficulty taking up ammonium through their roots. Thus, nitrogen-deficiency symptoms may appear in cold weather in plants growing in soils that have quite adequate total nitrogen. Nitrates *can* be absorbed by plants under cold conditions, but if plants are not growing rapidly enough to take up all available nitrates, these ions may leach downward beyond the reach of plant roots and contaminate the groundwater. Thus, it may be best to avoid application of nitrogen fertilizer to soils in late autumn, winter, or especially early spring, when plants are dormant.

The Phosphorus Cycle

By chemical analysis, most soils have enough total phosphorus to permit unrestricted plant growth. Why, then, are phosphorus deficiencies in plants so common? Simply put, most phosphorus in the environment is in highly insoluble or bound forms unavailable for uptake and use by plants and other garden organisms. In young or arid-region soils, phosphorus is mainly in the form of highly insoluble minerals such as apatite. In more weathered soils of humid regions, most phosphorus exists as fairly stable organic phosphates or as very insoluble phosphate salts of calcium and magnesium (in alkaline soils) or iron and aluminum (in acidic soils). In some tropical soils, large amounts of phosphates are tightly bound by clay. Soluble phosphates available for uptake by plants and most microbes generally constitute 1 percent or less of total soil phosphorus—and may be much lower at very alkaline pH.

Inputs into the garden phosphorus cycle include fertilizers, weathering of natural minerals, compost and manure, and certain detergents and pesticides. Once in the garden, most phosphorus-containing materials, whatever their original form, are rapidly converted into free phosphate ions by microbial enzymes. Free phosphate is then either absorbed by plants and microbes, or, more generally, precipitated as insoluble iron, aluminum, calcium, or magnesium salts. Inside of living organisms, phosphate is used to synthesize molecules such as DNA, RNA, the energy carrier ATP, and phytic acid (which is important in many seeds). When plants or other organisms die, their phosphate-containing molecules are either rapidly taken up and reused by other organisms; broken down to free phosphates that have the fates described above; or, less commonly, incorporated into stable organic phosphate polymers.

Factors that convert phosphates from insoluble or unavailable forms into available, soluble forms are very important to plant nutrition. To a limited extent, plants can mobilize phosphate by secreting various organic acids. However, by far the more significant mobilization of phosphorus is carried out by microbes (especially mycorrhizal fungi, discussed in chapter 5) that secrete organic phosphate-digesting enzymes. Mycorrhizae and other microbes can also release various acids and chelating agents that solubilize inorganic phosphates and keep them in solution until they can be taken up by living organisms.

Phosphorus in the garden ecosystem is quite immobile—certainly much less mobile than nitrogen, carbon, or oxygen—and losses by leaching to groundwater are slight (unless grossly excessive quantities of soluble fertilizers are applied). Losses of phosphorus to the atmosphere are almost nonexistent. Most phosphorus losses from gardens occur through surface soil erosion or through herbivory or harvesting of plants.

Excessive phosphorus in the environment is usually caused by human activities. When too much phosphate fertilizer is applied to soil, serious reduction in soil pH can occur with all its accompanying deleterious effects, such as solubilization of aluminum and other toxic substances. Worse, wastewater containing synthetic phosphate detergents, as well as runoff from overfertilized fields, lawns, and gardens, can contaminate streams and lakes. This contamination often causes a population explosion of aquatic microbes, especially algae, that may deplete the dissolved oxygen supply, causing water to become foul-smelling, murky, and unable to sustain fish or other higher forms of aquatic life (plate 72).

Cycles of Iron and Similar Elements

Iron and related elements such as copper, zinc, manganese, and cobalt, behave similarly in the garden environment. Thus, the cycle for iron can be considered more or less representative of the cycles for all of these plant nutrients. A particular characteristic of iron and similar elements is that they can readily and reversibly give up or gain one or more electrons per atom. Many soil microbes important in garden nutrient cycles take advantage of the energy released in this process and use these metals as energy sources—much like biological electrical batteries.

Iron is one of the most abundant elements in the Earth's crust; it is present in almost all soils in quantities far greater than are needed by plants. However, like phosphorus, iron in most soils is largely present in extremely insoluble forms unavailable for use by living organisms. The primary input of iron into garden ecosystems comes from weathering of naturally occurring primary minerals such as amphibole and pyroxene. These break down into secondary minerals such as iron oxides, which furnish the red and yellow colors that dominate soils of warm regions. Humans may also add iron in the form of discarded cans, tools, or other iron or steel items.

Most iron available for plant use is present in the soil as iron chelates formed by microbes. Chelates keep iron ions in solution, which enables them to be taken up by living organisms. Most iron in synthetic fertilizers is in the form of iron-EDTA, a synthetic chelate molecule. Inside living plants and microbes, iron is liberated from chelates and incorporated into a variety of molecules such as enzymes and cytochromes (important in cellular energy generation). After the death of an organism, iron-containing molecules may be metabolized by decomposer organisms; the iron is then either incorporated into the decomposers or released into the environment, where it is most often precipitated as insoluble iron compounds, thus continuing the cycle.

The pH and oxygen content of soils and waters have pronounced effects on iron availability to plants. Under neutral to alkaline pH and aerobic conditions, virtually all free iron is present as extremely insoluble salts. At strongly acidic soil pH, these iron salts become more water-soluble and thus more available for plants. Under anaerobic soil conditions there is a prevalence of forms of iron that tend to be much more water-soluble. These changes in water solubility of iron salts are reversible; that is, as soil pH or oxygen content changes in either direction, so also does the availability of iron. This pattern of pH- and oxygen-dependent changes in water solubility of iron and its availability for plants holds true for most related metals including zinc, copper, nickel, and manganese. Unfortunately, it is also true of a number of toxic heavy metals.

Iron and related metals tend to be quite immobile in the soil and are rarely lost by leaching. Most losses of iron from the garden occur via soil erosion or removal of plant material; however, there is generally so much total iron in soil that absolute losses are inconsequential compared to changes in availability. Metals such as zinc, molybdenum, and copper, on the other hand, are truly deficient in certain soils, and shortages may need to be remedied by fertilization as well as by appropriate management to enhance their availability for plants.

The Sulfur Cycle

Sources of sulfur in garden ecosystems include natural minerals (for instance, pyrites), gases (mostly in the form of acid rain), fertilizers, gypsum (calcium sulfate), compost, and mulch. Aluminum sulfate and flowers of sulfur are deliberately added by gardeners to lower soil pH for acid-loving plants such as blueberries and azaleas.

In aerobic soils and in natural waters, elemental sulfur and sulfur-containing minerals are rapidly oxidized to sulfates. Generally, sulfuric acid (hydrogen sulfate) is the immediate product of sulfur oxidation, so soil pH is lowered if sufficient alkaline materials, such as limestone, are not available to neutralize the acid. In the atmosphere, gaseous sulfur compounds such as hydrogen sulfide ("rotten egg" gas) and sulfur dioxide are oxidized to sulfur trioxide, which reacts with water to form sulfuric acid and falls to the ground as acid rain. Soluble sulfates are taken up by living organisms and transformed into molecules such as amino acids, vitamins, and even penicillin (by molds).

Under anaerobic conditions, sulfur compounds may be converted to sulfides that react with a number of metals to form very insoluble substances. Such compounds may also be metabolized by microorganisms to hydrogen sulfide, other foul-smelling gases, or elemental sulfur. The distinctive stinky smell and bluish gray or black colors of waterlogged soils, swamps, and marshes are mostly produced by sulfur-containing substances. When exposed to oxygen, as when wetland or compacted soils are dug up or drained, reduced sulfur substances are rapidly oxidized to sulfates. Large amounts of corrosive acids are often produced in the process—a major problem for construction projects in wetland or coastal soils. Oxidation and reduction of most forms of sulfur are readily reversible, so the anaerobic and aerobic sulfur cycles mesh together seamlessly.

Overall, sulfur is fairly stable and immobile in the soil, and deficiencies are much rarer than, for instance, nitrogen. Indeed, one irony of modern-day air pollution is that the large quantities of sulfur-containing gases produced by the burning of fossil fuels have made sulfur deficiencies all but unknown in developed areas of the world. To a limited extent, sulfur is lost from garden ecosystems, especially those with acidic soils, by leaching; but most losses are by soil erosion or removal of plant materials. Under anaerobic conditions, losses by volatilization of gaseous sulfur compounds can be significant, but few well-tended gardens are likely to have significant areas of such soils or waters.

Other Material Cycles

Virtually every substance of significance in gardens participates in a cycle. Potassium in gardens originates from weathering of natural minerals such as feldspar or greensand, or it may be added in the form of salts in fertilizers. In the soil, potassium may become a relatively stable component of various clay minerals, or it may exist as positively charged ions that can be reversibly bound to clay or humus, taken up by plants, or lost by leaching or runoff. Potassium in plant tissues generally persists as soluble ions used in a variety of life processes. Potassium deficiencies in gardens generally result from a deficiency of this element in the parent materials from which the soil was originally formed or from leaching of potassium downward out of the reach of plant roots resulting from high rainfall or irrigation water passing through acidic or sandy soil.

Calcium and magnesium in the garden are mostly found as minerals (for example, limestone) that are insoluble at neutral and alkaline pH. However, they also exist in free solution at acid pH and as water-soluble chelates or as positive ions reversibly bound to clay or humus. Once inside plants, calcium and magnesium tend to be immobilized as insoluble forms in cell walls or incorporated into molecules such as chlorophyll. When organisms die, their calcium and magnesium contents are liberated by decomposition; these elements are then taken up and reincorporated by other organisms, leached out of the soil, or precipitated as insoluble minerals, and the cycle begins anew. Most calcium and magnesium in gardens originates from soil parent materials. In the tropics, islands of relatively fertile soils are often found in regions where there are outcrops of limestone, basalt, or other alkaline rocks. New inputs of calcium or magnesium may come from addition of limestone (usually crushed before application) or dolomite (limestone high in magnesium rather than calcium); application of fertilizers such as calcium nitrate or Epsom salts (magnesium sulfate); or spillage of Portland cement or concrete.

So-called heavy metals such as arsenic, mercury, lead, and cadmium also go through complex cycles with many electrically charged states, organic forms, and often quite insoluble inorganic forms (especially at alkaline pH). These elements are often toxic to plants—and even more toxic to animals that eat plants containing them. However, the conversions from one form to another that take place in natural cycles may greatly decrease—or increase—heavy metals' toxicity. For instance, at one time waste metallic mercury was freely dumped into the environment in the belief that its virtual insolubility in water would make it nontoxic. Tragic poisoning incidents revealed that mercury-metabolizing bacteria, which no one had known existed, were capable of converting insoluble metallic mercury into water-soluble and extremely toxic methyl mercury.

Interconnections Among Cycles

It should be apparent that material cycles in a garden can themselves be viewed as subcycles of a grander, unified material cycle of the whole. Oxygen and carbon, especially, tend to tie together the subcycles; they make and break bonds with virtually every other element. A major objective of gardening should be to maintain the soil environment in an aerobic state (for example, by preventing soil compaction and providing good drainage), because virtually all products of the aerobic material cycles are more beneficial to plants than are those of the anaerobic cycles.

Living organisms also tie the cycles together. Within their cells, carbon, hydrogen, oxygen, nitrogen, sulfur, metals, and other elements are continually being bonded together into complex substances that can then be reused by other living organisms or degraded back to simpler chemical forms by decomposer organisms. The significance of oxygen-producing photosynthesis by plants, in particular, cannot be overstated. This important biological process converts chemically bound oxygen back into free molecular oxygen to continue the great interlocking material cycles of the garden ecosystem.

Limiting Factors

To produce a given amount of tissue, a plant may require a minimum amount of heat, light, water, nitrogen, or sulfur; a certain degree of freedom from herbivory; and so on. A plant will continue growing until any one of the resources required for growth becomes unavailable or until an inhibiting factor becomes too powerful. Then it will grow no more—no matter how much of the other needed resources are available. A critical resource or factor whose deficiency or excess limits the growth of the plant is known as a limiting factor. For example, a plant in a desert may grow in soil rich in mineral nutrients, with an abundance of light and heat, yet the plant may be stunted or even die for lack of water. In this case, water is the limiting factor. If the lack of water is relieved, the plant may grow larger until it encounters the next limiting factor, perhaps nitrogen deficiency. If nitrogen fertilizer is then added, the plant may grow yet more until it encounters the third limiting factor, perhaps insect herbivory, and so forth. Limiting factors for plant growth are most frequently shortages of water or available nitrogen, but could be almost anything else in shortage or harmful excess (plate 73).

It is important for gardeners to try to identify and, if possible, quantify the limiting factors in their gardens; these may vary from species to species, from season to season, or from spot to spot in the garden. Rigorously quantifying all factors that may limit plant growth in a garden may not be feasible—or even necessary. Some basic information, however, such as soil test results, simple home weather station data, and records of garden yields—combined with observations of such things as plant growth and development, sun/shade patterns, and insect populations—may enable the astute gardener to reasonably deduce which factor may most strongly limit plant productivity. For example, if poorly performing garden plants are in deep shade most of the day and the soil is normally rather damp, but soil tests show adequate levels of all critical nutrients, it would be reasonable to surmise that increasing sunlight exposure (for instance, by removing shading structures) would be more beneficial than applying fertilizer. Alternatively, one might decide to switch to growing shade-tolerant plants not limited by the existing environmental conditions.

Efforts should be focused on ameliorating limiting factors. It does no good—indeed it may do harm, both financially and environmentally—to pour on nitrogen fertilizer if, in fact, the limiting factor for a plant is insect predation, phosphorus deficiency, insufficient oxygen due to compacted soil, or . . . anything other than nitrogen deficiency.

Availability of energy may be the factor ultimately limiting productivity of a garden (or any other ecosystem), but material shortages or excesses are often more immediately obvious and amenable to human influence. Little can be done to prevent the ultimate loss of energy from an ecosystem, but gardeners can do a great deal to remedy nutrient deficiencies or imbalances, remove or neutralize toxic materials, or increase the efficiency of usage of energy during its sojourn in a garden. Ignorance of the laws of Nature is no excuse, and wise gardeners will learn and obey these rules. Roles of humans in the drama of the garden ecosystem are dealt with at length in chapter 6.

4

Plants in the Environment

Plants have been described allegorically as creatures that are stripped naked, buried up to their waists, and left in one place for years—totally at the mercy of their environment, unable to run away, fight back, or multiply faster than their adversaries. However, as this chapter will make clear, in reality, plants have a remarkable ability to adapt to their respective environments.

Interactions of Plants with Natural Forces

Heat, light, gravity, motion, pressure, and other forces profoundly influence plant growth and development. Plants, in turn, can modify these physical forces in the garden environment so as to either foster or hinder their own growth and activities as well as those of other garden dwellers.

Heat

The temperature of the air, soil, and water surrounding plants has dramatic effects on most aspects of plant growth and development. However, with rare exceptions such as the heat-generating western and eastern skunk cabbages (*Lysichiton* and *Symplocarpus foetidus*, respectively; plate 74) and lotus (*Nelumbo*), plants are unable to alter or regulate their own internal temperatures. Therefore, plants are largely at the mercy of the temperature of their environments.

Cold air and soil slow plant growth and development, and, if severe enough, cause chilling injury or freezing. Conversely, warmth may accelerate plant growth and development, up to a point. Excessively high temperatures may inhibit growth by accelerating respiration more vigorously than photosynthesis, resulting in exhaustion of food reserves and loss of vigor. High temperatures may also damage plant proteins required for metabolism. Also, many biochemical substances that impart flavor or aroma are volatile oils that may evaporate and dissipate at high temperatures. Thus, for the most flavorful herbs or the most aromatic flowers, it is best to harvest and store them in relatively cool conditions, then warm them for eating or smelling.

A rather arbitrary, but nevertheless valuable horticultural categorization is cool-season versus warm-season annual crops—those plants that grow optimally at a temperature below about 68°F (20°C) versus those that grow optimally above that temperature. Cool-season plants include most leafy greens such as lettuce, spinach (*Spinacia oleracea*), and endive (*Cichorium endivia*); many root crops such as carrots, parsnips (*Pastinaca sativa*), and salsify (*Tragopogon porrifolius*); cole crops (family Brassicaceae or Cruciferae); and some delicate flowers such as mignonettes (*Reseda odorata*), violets, delphiniums, and annual larkspurs (*Consolida ajacis*). Warm-season plants include most annual flowering and fruiting plants, such as zinnias (*Zinnia*), cucurbits, tomatoes, peppers, and eggplants/aubergines.

Most temperate-zone plants require higher temperatures for seed germination than for seedling growth and higher temperatures for seedling growth than for growth of maturing plants, although lettuce, for example, is an exception. Unfortunately for gardeners in most temperate areas, spring soil temperatures follow exactly the opposite pattern—starting out cold, then warming. In early spring, slowly germinating seeds often languish in frigid soil, leaking nutrients that attract cold-tolerant, seed-rotting soil fungi. When survivors finally warm enough to germinate, the soil temperature may climb above the optimum for vegetative growth. This is one reason why many cool-season crops can be more successfully grown in autumn—a season when many weed-weary gardeners are checking in their hoes and welcoming frost to relieve their labors. To optimize plant production, professional growers who use greenhouses often maintain warm chambers for germinating seeds, then step down emerged seedlings through progressively cooler chambers until they reach a lower optimal growing temperature. Home gardeners can germinate seeds in warm spots such as gas ovens with pilot lights, above refrigerators, on warming mats, or in south-facing windows and then move the seedlings to cooler locations. Whatever a gardener's strategy, seeds should never be sown in soil below the minimum germination temperature for that plant. To do so is a double sin: not only will the seed almost never germinate, and thus be wasted, but populations of seed-rotting soil fungi will increase and lie in wait for the next batch of seed.

Though warmth may be needed for germination, many seeds need prior stratification (cold treatment), especially those of temperate-zone perennials. Cold stratification appears to be a natural way of delaying seed germination from autumn until spring—to prevent sprouting just before the bitter cold of winter. Double stratification delays germination of part of a seed crop until the second spring, thus spreading out germination over two years. This trait may be advantageous for survival of a plant population because only a fraction of a given seed crop will be at risk during a particular growing season. Such seed dormancy and staggered germination, however, may be very inconvenient for gardeners; thus, the trait has been bred out of most domesticated plants.

Maturing plants are also affected by low temperatures. Spells of chilly weather may trigger premature flowering and seed set, known as bolting, in plants such as Chinese cabbage (*Brassica chinensis*), beets, and celery (*Apium graveolens*). Low nighttime temperatures may cause many warm-season fruiting crops, such as tomatoes and peppers, to drop flowers or it may prevent flowering altogether. Low temperatures also generally inhibit fruit ripening. For example, most tomatoes will not ripen below 50°F (10°C).

Low soil temperatures may cause deficiencies of nutrients and, perhaps surprisingly, of water. Warm-season plants grown at suboptimal temperatures are especially susceptible to nitrogen and phosphorus deficiencies. Cold-induced nutrient deficiencies seem to result both from indirect and direct effects. Indirectly, low temperatures may inhibit activities of soil microbes that convert soil nutrients from unavailable forms into those more available to plants. Directly, temperature also strongly affects the abilities of roots to absorb nutrients and water, primarily because of biochemical and biophysical changes in root cell membranes. Differences in these effects seem to be one of the factors most clearly distinguishing cool-season and warm-season crops. For instance, roots of watermelon (*Citrullus lanatus*), a warm-season crop, can absorb water only 6 percent as efficiently at 54°F (12°C) as at 77°F (25°C), whereas water uptake by collards (*Brassica oleracea* Acephala Group), a cool-season crop, is virtually unaffected at 54°F compared to 77°F, and at 43°F (6°C) it can still absorb water 63 percent as well as at 77°F.

Water temperature also affects plant growth and development. Plants no more benefit from being scalded or plunged into ice water than we do! They can actually suffer a form of shock, with consequent growth retardation or even permanent stunting, if abruptly exposed to water at extreme temperatures. Cold water chills soil much more strongly than does cold air. Gardeners have little control over the temperature of rainwater, but cold tap water can be warmed by running it through a long, dark-colored hose or pipe exposed to sunlight. Water used to irrigate houseplants, transplants, or, especially, young seedlings growing in containers should be allowed to come to room temperature before use. Boiling water can be used to kill plants—even such hardy woody perennials as poison ivy (*Toxicodendron radicans*)—although repeated applications may be necessary to complete the job.

For many crops, maturation can be fairly accurately predicted by accumulation of degree-days (sometimes called heat-units). Degree-days are calculated by recording at a standard time each day the number of degrees by which the air temperature exceeds a crop-specific threshold temperature. The degree-days are then summed from the date of planting, transplanting, or bloom of flowers as applicable. If temperatures fall below threshold values, rather than recording negative values, no degree-days are recorded. The crop will reach a given state of maturity when it accumulates a certain total of heat-units. Tasseling of sweet corn can be especially accurately predicted, as can date of bloom, fruit development, and maturity of some commercially important fruit crops and blossoming of some ornamental flowers. Unfortunately, accurately calculated degree-day maturation values are available for only a few commercially important plants, and these are frequently found in technical or academic publications or in proprietary literature distributed by vendors of plants or seeds to their customers.

Seed catalogs often provide days to maturity of various plant varieties. Many exasperated gardeners have found these figures wildly inaccurate and of real value only for roughly estimating sequence of maturation of different varieties. Seed catalogs would be much more accurate if they gave heat-units to maturity. Why don't seed companies give out these numbers for home gardeners to use? One reason: they're generally not known for less common or noncommercial crops, and they often vary greatly from variety to variety in even well-known crops. Also, most gardeners are unlikely to keep and analyze the detailed weather records necessary to apply these values to their gardens. So, the often fictitious days-to-maturity figures continue to be used, although it's worth keeping in mind that what really matures plants is not passage of time per se, but accumulation of energy in the form of heat.

Most temperate-zone perennial plants, including most deciduous North American and Eurasian fruit and nut trees, enter a winter dormancy. These plants remain dormant, even if exposed to warm temperatures, until they have satisfied their chilling requirement, or, in horticultural terms, have become vernalized (plate 75). Chilling requirements are measured in chilling-units (analogous to heat-units), usually degree-days or degree-hours below a certain critical temperature. For example, in most peach varieties the critical temperature is about 45°F (7°C) and chilling-units are defined as hours below that critical temperature after reaching full dormancy. Thus, serious peach growers use recording thermometers to sum up hours spent by their trees below 45°F from full leaf-drop onward to forecast timing of such developmental events as bloom and leaf emergence. North American peach varieties have chilling requirements ranging from 200 to 1250 hours, with most varieties in the 800- to 1000-hour range. Some hardy central Asian varieties with much higher chilling requirements are being considered as breeding stock for producing hardier peach trees (although problems with seed-borne Asian fruit tree viruses have hampered the introduction of breeding stock). Once a peach tree has satisfied its chilling requirement, the flower buds will open with the first warm weather. A little arithmetic will show why most peach varieties cannot be successfully grown in mild-winter areas like the Gulf Coast—there are simply not 800-plus hours below 45°F in the average winter. Only low-chill peach varieties can be grown in such regions. On the other hand, gardeners in severe winter areas need high-chill peach varieties to delay bloom as long as possible in the spring so the trees do not blossom at the first thaw.

Many horticulturists somewhat arbitrarily—but usefully—divide annual plants into four cold air–tolerance classes (plate 76). Very tender plants are killed or severely injured by above-freezing chilling or brief exposure to even light frost (32°F, 0°C) and include most cucurbits, zebrinas or spiderworts (*Tradescantia*), portulacas (or moss roses, *Portulaca grandiflora*), and most tropical houseplants. Tender plants are injured by light frost, but may survive a brief exposure, and include most peppers, tomatoes, sweet corn, and zinnias. Half-hardy plants can survive but are injured by a brief killing frost (28°F, –2°C) and include beets, carrots, celery, and lettuce. Hardy plants are able to survive and grow at temperatures below 28°F (–2°C) and include most cole crops, alliums such as onions and garlic, spinach, chrysanthemums (*Chrysanthemum*), and pansies (*Viola* ×*wittrockiana*). However, even hardy species can be injured or killed if abruptly transferred from mild to subfreezing temperatures without an intervening period of exposure to cool temperatures to activate cold-tolerance mechanisms. Once adapted, however, some hardy annual plants are amazingly cold-tolerant. Some varieties of kale (*Brassica oleracea* Acephala Group), mustard (*Brassica juncea*), and Brussels sprouts (*Brassica oleracea* Gemmifera Group) can survive temperatures below 0°F (–18°C).

Most temperate-zone annual plants do not seem to be directly injured by the slowing down of biochemical processes at low temperatures, but indirectly by ice crystals forming within their tissues. The reason for widely defining a killing frost as 28°F (–2°C), rather than the freezing point of pure water, 32°F (0°C), is that many plant tissues contain dissolved sugars, salts, and other substances, which lower the freezing point of the cell sap to 28°F or below. Even many frost-sensitive plants can be experimentally supercooled well below 28°F; as long as ice crystals do not form, they can often revive without apparent injury when rewarmed. However, the presence of dust particles or certain surface-dwelling ice-plus bacteria can trigger ice crystallization at temperatures as high as 39°F (4°C) with consequent frost injury to plants. There are now commercially available preparations of ice-minus bacteria that can be applied to plant surfaces to overwhelm ice-plus bacteria and thus reduce frost injury.

One way most hardy annual plants adapt for growth at subfreezing temperatures appears rather straightforward: they accumulate high levels of dissolved sugars and sugar alcohols in their cell sap, which act effectively as antifreeze. The fact that most of these substances are also sweet-tasting may explain the long-standing observation that cold-hardy Brussels sprouts, turnips, and parsnips, for instance, often taste better after a freeze than before. More subtle biochemical adaptations to cold, such as alterations in cell membranes and enzymes, may also exist, but these have been less investigated.

The growing season for most temperate-zone annual plants can be calculated by counting the number of days between the average date of the last killing frost in the spring and the average date of the first killing frost in autumn. Using climatic data for their areas (available from the U.S. Department of Agriculture or the U.S. Weather Service, Environment Canada, the Royal Meteorological Society in the United Kingdom, or equivalent agencies in other countries), gardeners in different areas can choose varieties and species that have a high probability of maturing in this frost-free time span. The cold survival of perennial plants, however, is better predicted by minimum winter temperatures than by frost dates or average seasonal temperatures. After all, a perennial plant will be killed by extreme, not average, temperatures. The U.S. Department of Agriculture has used local weather records from all over North America to produce a periodically updated hardiness zone map that is almost universally cited in nursery catalogs to aid gardeners in selecting perennial plant species and varieties (plate 77).

Winter-kill, the injury or death of perennial plants caused by severe cold, is a bit of a mystery. We do not really understand why a given plant might survive at, say, −20°F (−29°C), but not at −25°F (−32°C). After all, ice crystals probably form in the plant's tissues well above either temperature. Also, the plant is dormant at either temperature, so one might expect few if any biochemical differences between the two temperatures. However, the absolute humidity of air decreases sharply with declining temperature, and one postulated explanation of winter-kill is that it is due to dehydration of plant tissues, with some plants resisting freeze-drying better than others. It is also well known that trees that have been stressed during the growing season by drought, disease, or insect attacks are more subject to winter-kill than those that have not. However, a flush of vigorous growth resulting from nitrogen fertilization or excessive watering during late summer or autumn also reduces winter hardiness, so the ultimate mechanism of winter-kill is still quite unclear.

In contrast to temperate plants, many truly tropical plants suffer chilling injury or are killed by temperatures well above freezing. Most species of bananas, for instance, are injured by temperatures below 50°F (10°C) and, outside of the tropics, the plants must be overwintered indoors (plate 78). The physiological mechanism of this above-freezing, low-temperature intolerance is not well understood but has significant consequences for would-be growers of tropical plants—and for houseplant lovers during winter power outages.

It is clear that perennial plants have distinctive minimum temperatures below which they sustain serious injury or die, even when apparently fully dormant. And, like the hardening of annuals to frost, the adaptation of perennials to cold requires conditioning, with gradual exposure to lower temperatures. In 1991 there was an exceptionally mild autumn in Missouri; temperatures were around 70°F (21°C) on October 31, but fell the next day to as low as −15°F (−26°C). Many unhardened woody perennials literally exploded as their sap rapidly froze and expanded (plate 79). Some agricultural authorities estimated that half the peach trees in Missouri died that day! As this dramatic example illustrates, moisture and temperature strongly interact in their effects on plants.

In temperate regions, temperatures are usually highest during seasons when most plants are actively growing and reproducing. Some plants, though, especially cool-season grasses such as fescues, bluegrass (*Poa*), and ryegrass (*Lolium*), escape heat stress by going dormant during the heat of the summer. Midsummer browning of American lawns, which consist predominantly of cool-season grasses, is biologically normal; the grasses will recover and green up again in cooler autumn temperatures. When homeowners and others pour on water and fertilizers to prevent this natural dormancy and have the greenest lawn on the block all summer long, they subject the grasses to physiological stress. Although lawns treated thus may appear lush—and require frequent mowing—they may actually be weakened and require increased applications of herbicides and pesticides. There are, of course, warm-season grasses, such as zoysia (*Zoysia matrella*), Bermuda grass (*Cynodon dactylon*), and many native prairie grasses such as big bluestem (*Andropogon gerardii*) that are naturally green in the summer and brown in the spring and autumn. However, because of their coarse texture, clumped growth habit, and often dull appearance (which are, in part, results of adaptation to heat and drought), warm-season grasses are generally unpopular for lawns outside of hot southern areas, where cool-season grasses may not grow well at all.

Actively growing plants vary widely in their tolerance of and adaptation to high temperatures. When cool-season plants are exposed to excessively warm temperatures, they may cope, but the results are generally undesirable from a gardener's point of view. Plants may become stunted or deformed; stems, roots, and leaves become tough and woody or leathery; pungent-tasting substances accumulate in tissues; and flowers become bleached, burned, or drop off altogether. Of course, warm-season plants that are exposed to extremely high temperatures may respond as do heat-stressed cool-season plants, although they generally show greater ability to survive excessive heat without irreversible damage.

In virtually all plants, however, one process is especially sensitive to excessively high temperature—sexual reproduction. Pollen grains frequently become sterile and fertilized ovules may abort at temperatures well below those at which vegetative tissues suffer gross harm. Most tomato varieties, for instance, cannot set fruit above about 92°F (33°C) daytime temperatures or 77°F (25°C) nighttime temperatures, although the plants themselves can tolerate much higher temperatures if adequate water is present. The American Horticultural Society has produced a commercially available heat zone map of the United States that shows areas with similar average number of days per year in which temperatures reach 86°F (30°C) or above. This map will assist gardeners in selecting plants that can survive and grow in their local summer heat.

Light

One of the most obvious and distinctive features of green plants is that, well . . . they're green. The nearly ubiquitous green of plants is due to chlorophyll, a group of pigments that absorbs light energy used to drive the critical process of photosynthesis. Both the quantity and quality of light greatly affect the functioning of plants in dramatic and subtle ways. Without light, plants cannot photosynthesize. But plants neither die nor go dormant during brief dark periods, such as at night. In both light and dark, plants can use sugars, fats, and other organic molecules as sources of energy and materials for synthesis of tissues, just as animals do. Thus, in every plant there is a balance between creating new sugars (photosynthesis) and burning them (respiration).

As the intensity of light to which a plant is exposed increases, so does the rate of photosynthesis—up to a light level called the light saturation point, beyond which there is no increase in photosynthesis. At low light intensities, respiration may exceed the rate of photosynthesis, so the plant must draw on reserves of previously photosynthesized carbohydrates. The intensity of light above which photosynthesis of sugars exceeds their breakdown by respiration—so that net growth occurs—is known as the light compensation point. Compensation points of plants vary widely, as do saturation points. Some deep woodland shade plants such as caladiums, trilliums, coleus, hostas, philodendrons, and many ferns have very low compensation points and can grow at low light intensities; but they also have low saturation points and may be injured by bright light. Not surprisingly, many common houseplants, able to grow as they do in constantly warm but relatively dim conditions, have origins as deep-shade, tropical forest plants. Tropical understory plants such as cacao, coffee (*Coffea arabica*), and tea grow poorly in full sunlight and generally require shading. Desert plants such as euphorbs and cacti and warm-season grasses such as corn/maize, sugar cane, and zoysia, have high light compensation and saturation points and can use light up to a very high intensity—but suffer in shade or cloudy weather. Even plants that have high light saturation points may receive more light than they can use, however, especially on cloudless tropical or midlatitude summer days. Under these conditions, photosynthetic rates may level off as the saturation point is exceeded, while respiration increases due to photorespiration. At very high light intensities, the result can be a surprising net decline in synthesis of sugars and consequent plant debilitation.

Low-light injury to plants is usually a gradual process. With progressive exhaustion of energy reserves due to respiration exceeding photosynthesis, plants may become pale, spindly, twisted, and stretched out (etiolated), with weak, brittle, watery tissues (plate 80). Light-starved plants usually have low disease resistance, and pathogenic microbes may do more serious damage than usual. Etiolation of light-deprived plants may be an adaptive response as the plants reach toward the nearest light source. The wider spacing of leaves on etiolated plants also reduces mutual shading and exposes stems, thus maximizing the amount of light striking the photosynthetic plant tissues. Furthermore, pores of light-deficient plants open wide to admit carbon dioxide freely and enhance photosynthesis as much as possible. Gradual re-exposure to adequate light will stimulate etiolated plants to synthesize more chlorophyll and green up again. The spindliness is usually not reversible, however, and the plants may be misshapen for life, although judicious pruning or deep replanting to partially bury stretched-out stems may help to correct some of the deformation. Etiolation, whether in houseplants or in garden plants, is a message to the grower that more light is needed. If feasible, light intensity may be increased by adding stronger artificial lighting (indoors), by removing objects that cast shade, or by moving plants to a more favorable site.

High-light injury can occur very quickly. When exposed to bright light, especially if suddenly, shade-adapted species or etiolated plants of any species may suffer rapid dehydration, bleaching, and various forms of sunburn. Especially at risk are fruits, flowers, and some stems that have little chlorophyll. Even in plants adapted to bright sunlight, sunburn is a common problem for fruits such as apples, peppers, tomatoes, peaches, and beans (plate 81). Indeed, many seed catalogs prominently feature good leaf cover as a positive attribute of varieties of fruit-bearing plants; these varieties cover their fruits well with leaves to protect them from sunburn. Some ornamental flowers, including zinnias (especially the green-flowered varieties), bells-of-Ireland (*Moluccella laevis*), and gladiolus (*Gladiolus*), suffer from sun-bleaching of flowers—although the plants themselves grow best in full sun. Trunks and limbs of perennial woody plants are often sunburned during winter, when the plants are defoliated, especially on southwest-facing slopes. Whitewashing tree trunks in autumn is a time-honored technique used by nurseries and orchards for reducing this injury.

To a certain extent, plants can adapt to excessive light intensity if exposed gradually. This can allow plants to become darker green as more protective chlorophyll and other pigments are made. Plants may also become squatter as newly formed internodes elongate less than under low-light conditions, causing branches and leaves to be spaced more closely together. Stems and leaves may also become thicker and tougher with increased woody and waxy layers, and the stomates may close or contract to reduce moisture loss.

The color of light also has a major influence on plant growth and development. Chlorophyll absorbs red and blue light most efficiently. Some yellow and orange plant pigments (such as carotenoids) absorb violet and blue light and contribute to photosynthesis as well. No photosynthetic pigments significantly absorb green or yellow green light, though Place a plant under light of those colors, no matter how intense, and it will develop symptoms of low-light injury. Of course, gardeners can do little about the color of sunlight, which tends to be fairly even across the visible spectrum except for a slight enrichment in yellow. Except in deep shade or heavily overcast conditions, however, sunlight intensity is usually well above the compensation point of most plants at all wavelengths (colors) and is thus not limiting for photosynthesis.

In addition to photosynthesis, other plant processes are strongly affected by exposure to light of particular colors—especially blue, red, and far-red (wavelengths just beyond visible red). In most plants, for example, light-induced opening of stomates and phototropism (the tendency to grow toward or away from a light source) are mediated by blue wavelengths. In contrast, day-length effects on plant flowering and other developmental events are caused by red and far-red light. Plant sensitivity to light of particular colors may relate to environmental conditions. Blue sunlight is most intense when sun angle is high (midday or near the summer solstice). Sunlight is enriched in red at low sun angle (near sunrise or sunset or in midwinter), although this apparent color change is mainly due to relative reduction in blue and violet light rather than to intensification of red wavelengths. Perhaps shifts in the relative intensity of blue and red light mediate plant processes. These effects of light color, of course, are generally beyond the outdoor gardener's control. But nearby artificial nighttime light sources such as high-intensity streetlights and security lights have been known to disrupt plant developmental cycles both by providing an unbalanced color spectrum and by altering day-night cycles. In some cases, removal of the offending light source, shading, or relocation of the affected plants may be necessary.

Colors of indoor artificial lights are another story. For example, tungsten bulbs are rich in the red-to-yellow range and low in blue-violet, whereas cool-white fluorescent bulbs are the opposite. This is the reason for the development of expensive grow lights, fluorescent bulbs specially manufactured to mimic the color spectrum of sunlight. Most houseplants and transplants started indoors, however, will grow acceptably well under the less expensive cool-white fluorescent lights—provided that the total intensity is high enough (such as that given by fresh bulbs hung close to the plants). Apparent exceptions are African violets (*Saintpaulia*), succulents, roses, and cacti, which really do seem to benefit significantly from the sunlight-imitation spectrum of grow lights, especially for blooming. Although indoor lighting can approximate the color spectrum of natural sunlight, no artificial light source commonly used by gardeners approaches sunlight in intensity. Virtually all plants grown indoors, except for tropical houseplants adapted to deep rainforest shade, will experience low-light injury to some degree.

Plants respond to the direction of light sources (phototropism) as well as to the light energy itself. They will generally grow toward any strongly directional light source, especially if stressed by low light intensity (etiolation being an extreme case of stretching toward light). Witness two trees planted too close together; they will often grow away from each other, both striving to maximize sunlight exposure. Orientation of seedlings and their leaves toward the outdoors is common in plants grown in front of windows. To keep windowsill plants from developing a permanent lean, rotate them every few days. Bear in mind, though, that leaning or stretching plants may be trying to tell you that they are suffering from inadequate light.

Plants also actively reorient leaves during the day, tracking the sun across the sky from east to west. This particular phototropism is exhibited by many plants, even large trees such as maples and elms (*Ulmus*), but is most conspicuous and best known in the aptly named sunflower. Young sunflower plants rotate their main stems during the day to track the sun, then reset themselves at night so their solar collectors (leaves) are aimed at the rising sun in the morning. Another well-known example of solar tracking is seen in mountain avens (*Dryas*), a member of the rose family found in the Arctic. As their parabola-shaped flowers follow the sun through the day, insects take advantage of this opportunity to sun themselves within the flowers.

Plants are also strongly influenced by the duration and timing of light exposure. A dramatic effect of day length is to induce or prevent flowering or other developmental events. Some plants are known as short-day plants, meaning that they flower when day length declines below a certain critical duration, usually in late summer or autumn. Long-day plants, in contrast, flower when the light period surpasses a certain critical length, usually as spring approaches summer. Yet other plants are day-neutral; they flower when they reach a certain stage of physiological maturity regardless of the daily light period. Examples of day-neutral plants include most tomatoes, cucumbers, gladiolus, and grasses.

Day-length effects have major consequences for gardeners and farmers. Cole crops, spinach, lettuce, most strawberries, and spring bulbs are long-day plants; they naturally bloom when days reach and exceed certain lengths. To a gardener, this may be viewed as good (blooming of tulips, narcissus, crocus, and strawberries) or bad (bolting of lettuce, spinach, and spring-planted cabbage). Most annual weeds are short-day plants that bloom and set seed as day length declines after midsummer (very, very bad), as do beans and chrysanthemums (good). North of about 38°N latitude, southern varieties of sweet onions such as 'Granex,' and 'Texas Sweet' tend to prematurely form only dwarfed bulbs due to excessively long summer days. In contrast, northern onion varieties such as sweet 'Walla Walla' or pungent 'Spartan' rarely form bulbs at all in southern areas for the opposite reason. Gardeners can do little about the timing of sunrise and sunset but can attempt to compensate for day-length effects by choosing species or varieties that are relatively insensitive to day length (such as 'Candy' onions or 'Tribute' strawberries). But it's still much easier to raise tulips, potatoes, and strawberries in spring to early summer, and cabbage, spinach, radishes, and chrysanthemums in the late summer and autumn—during their natural day-length-sensitive blooming or nonblooming periods, depending on whether or not blooming is desired.

The term *day length* has become fixed in the literature, so we use it in this text. In reality, however, plant development mechanisms sense the length of dark periods (nights), not day length. So short-day plants are actually "long-night" plants and long-day plants are "short-night" plants. The fact that apparent day-length effects are really night-length effects may have little relevance to the rural gardener with few outdoor light sources other than the sun, but it can strongly impact urban and indoor plant-raisers. Interrupting the dark period of sensitive plants with a streetlight, security light, room lights, or even a vehicle's headlights may make plants sense that the days are very long (nights very short) indeed. For some plants, such as day-neutral houseplants of tropical rainforest origins, this may make little difference. But what about plants such as cabbage that naturally flower when nights get short in late spring and early summer? Well, it depends if you want your indoor-grown cabbage transplants to produce heads or flowers.

Another consequence of actual day length is that accumulation of light energy for photosynthesis depends both on intensity and duration of exposure. Gardeners who raise plants indoors or in greenhouses during winter may need to increase effective day length and compensate for the relative dimness of winter sunlight with supplemental artificial lighting.

Slower growth and development of outdoor plants in autumn is a result both of cooler temperatures and reduced incoming light energy due to shorter days as well as the lower angle of the sun in the sky.

Light, like temperature, can influence seed dormancy and germination. Seeds of coleus, African violet, and impatiens, for instance, only germinate well in the light, whereas echinacea, cyclamens, and nasturtiums (*Tropaeolum*) require darkness. Then again, seeds of many plants are light-indifferent, germinating whenever moisture and temperature are favorable. Some plants even shift from requiring dark to requiring light if the seeds are buried before they have been exposed to proper temperature or moisture conditions. Light sensitivity of seeds may function like stratification by delaying seed germination until conditions are favorable or spreading seed germination over an extended period to hedge bets.

Light interacts with other factors to affect the growth of plants. For instance, critical dark periods for day-length-sensitive plants are somewhat different at different temperatures; cold-exposed spring-planted cole crops will bolt at shorter day lengths than those maintained at warmer temperatures. Nitrogen nutrition (too much or too little) may also impact a plant's response to high- or low-light conditions. Generally, high levels of nitrogen will tend to keep a plant in a vegetative state; low levels tend to stimulate early flowering.

Other Electromagnetic Radiation and Magnetism

Visible light is only a small portion of the spectrum of electromagnetic energy, which ranges from long-wavelength, low-powered radio waves at one extreme to energetic, short-wavelength gamma rays at the other. Plants are exposed to natural and artificial long-wavelength electromagnetic radiation, including microwaves and infrared radiation. Despite many experiments and anecdotal accounts, it is not clear that long-wavelength radiation has any significant biological effects on plants at the intensities normally encountered in the environment.

More energetic short-wavelength radiation, ranging from ultraviolet to gamma rays, can have pronounced effects, however. Plants are generally more sensitive than animals to sunburn from ultraviolet light, because ultraviolet rays can essentially burn out the photosynthetic machinery, much as an electrical power surge can short out a sensitive electronic circuit. Thus, an increase in average intensity of ultraviolet light reaching the

Earth, such as may occur with the thinning of the ozone layer in the stratosphere, could have negative effects on plant health and productivity. Some insect-pollinated plants, however, such as marsh marigolds (*Caltha*), cinquefoils (*Potentilla*), carnations and pinks (*Dianthus*), and periwinkles (*Vinca*) benefit from low-intensity ultraviolet light that illuminates pigment patterns on their flower petals. These patterns form pollination targets or nectar guides for insects such as honey bees, which can see ultraviolet light. (These targets can be viewed by enterprising photographers through ultraviolet light photography.) Effects on plants from high-energy ionizing radiation, such as gamma rays from radioactive substances, are not well understood. What limited evidence there is, however, suggests that plants are considerably more resistant to obvious damage from high-energy radiation than are most animals.

There is considerable interest in magnetism as a treatment for human ailments such as rheumatism, and many people have taken to wearing magnetic bracelets and other paraphernalia. Magnetic fields can strongly affect some species of migratory birds and bacteria—including some soil dwellers. However, no significant effects of magnetic fields on plants have been convincingly demonstrated, despite much folklore and repeated scientific investigations.

Vibration

Can talking or singing to your plants really help them grow better? The jury is still out on this popular question, but it is known that plants are affected by vibration, of which sound is only one type. Excessive vibration, such as violent shaking, can cause stress responses in plants and stunt them or shorten their lives. Modern commercial harvesting of English walnuts (*Juglans regia*), pecans (*Carya illinoinensis*), almonds (*Amygdalus communis*), and other nuts has come to depend on mechanical shaking of trees to drop as many nuts as possible at one time into collection devices. Trees harvested in this fashion have significantly shorter productive lifetimes than do trees that are not shaken but harvested by hand the old-fashioned (much more expensive) way. Several studies have been carried out to establish the best frequency and amplitude of shaking to optimize the trade-off between short-term gains from reduced harvesting costs and long-term losses from tree decline and premature death.

Although violent shaking of plants is harmful, moderate vibration or swaying of plants—such as by the wind—may have some positive effects.

Newly transplanted trees may tip over or lean away from prevailing winds before their root systems become established and their trunks thicken, so it became a popular practice some decades ago to firmly fasten young trees to posts so they would grow straight. It was eventually discovered, however, that when such trees eventually outgrew their supports and had to be released, they were weak and subject to catastrophic breakage in high winds or during snowstorms. It seems that the alternating compression and extension of trunks and limbs while swaying stimulates the production of fibers and elongated cells that give wood its tensile strength. Currently accepted nursery practice is a compromise: trees are still staked to keep them from tipping over, but they are loosely secured with elastic ties to permit a limited range of wind-caused swaying and consequent wood-strengthening processes (plate 82).

Gravity

Plants are very sensitive to spatial orientation and exhibit a wide range of geotropisms (movements or growth in response to gravity). Roots normally grow down and shoots normally grow up, no matter how a seed is placed in the soil. However, roots and shoots also emerge from particular locations on a seed or bulb, regardless of its orientation. Thus, if germinating seeds are randomly positioned in a growing medium, some roots will emerge below seeds and be properly oriented to grow immediately downward. Others will emerge sideways and need to make a 90° turn, while still others will emerge on top and must get turned completely around. The same thing occurs in reverse for shoots. Does seed orientation matter? Experiments say, yes, it does. Plants that must get turned around may start life with a handicap, and, other things being equal, may never catch up with better-positioned peers. For most small-seeded plants this positioning effect is weak. But for large-seeded plants such as beans, the effect is stronger. For plants grown from bulbs or corms, including gladiolus, tulips, hyacinths (*Hyacinthus*), onions, and garlic, positional effects at planting may be critical. An upside-down gladiolus bulb may not grow at all, or at least the shoot may never emerge from the soil.

Researchers have played a dirty trick on plants by growing them in a spinning centrifuge. Under some conditions, plants mistake the centrifugal force for gravity and grow sideways or even upside down, depending on the angle of the plane of rotation of the centrifuge. Of course, few gardens are constructed on a centrifuge, although one famous English garden does have a flower bed on a slowly rotating gigantic clock dial. More naturally, if a plant is tilted or bent over, such as by the wind or an animal, the younger, still flexible, aboveground parts of the plant may exhibit geotropism by bending upward in response to gravity.

Pressure and Constriction

Other forces affecting plants include pressure, constriction, and impact. If a plant is constrained by a mechanical device or by a natural obstacle, such as another plant or rock, it will grow so as to bypass the obstacle, even if it has to overrule its natural geotropic or phototropic responses. Growth response to touch or pressure is called thigmotropism. Bonsai fanciers make use of this plant characteristic to train their oddly distorted miniature trees. By far the greatest effect of pressure on plants, however, is exerted indirectly via soil compaction. Plant roots may face intractable obstacles in growing through dense soil layers formed by compaction or deposition of mineral salts or clay, and plant growth may be severely constricted. Thus, plants left too long in pots may become rootbound, a condition that often includes a thigmotropic response of roots circling about the inner pot face. These circling roots must be pruned or disentangled when rootbound plants are transferred to more spacious surroundings, or the plants may never satisfactorily take root. Another thigmotropic response is illustrated by such plants as the sensitive plant (*Mimosa pudica*), which folds up its leaflets in response to touch, or by carnivorous plants such as the Venus flytrap, which rapidly closes its specialized leaves to entrap insects that contact touch-sensitive surfaces or hairs.

If a wire or other band is fastened too tightly about a plant's stem, it may constrict the flow of fluids in the phloem just under the surface, thus preventing movement of sugars from leaves to roots and causing the eventual decline and demise of the plant. (The deeper xylem will probably not be affected as much, so water and mineral nutrients may continue to move from roots to leaves until the roots deteriorate.) In the short term, however, such a tourniquet will cause accumulation of sugars and hormones in the leaves and stems, leading to such effects as enhanced flowering and fruit set. This phenomenon may be taken advantage of by fruit tree growers to shock recalcitrant young trees into initiating blooming and fruit-bearing— although great care must be taken not to leave the tourniquet on so long as to seriously injure or kill the tree. If a constricting band or wire is tied only partway around a stem or trunk, or if it is loose in some places, plants may attempt to compensate for partial strangulation by increased growth (hypertrophy) of phloem tissue at places where constriction is incomplete.

Interactions of Plants with Mineral Substances

Like all other organisms, plants need a variety of chemical substances as building blocks for their tissues or as sources of energy. Plants are capable of synthesizing complex molecules from simpler precursors, however, and their needs can be met by a menu of relatively basic substances. Hydrogen, oxygen, carbon, nitrogen, phosphorus, potassium, and calcium are required in fairly large quantities (hundreds to thousands of pounds per acre or kilograms per hectare). These staple foods are collectively termed *macronutrients*. In contrast, plants need the mesonutrients iron, magnesium, and sulfur in much smaller amounts (a few pounds per acre or kilograms per hectare). The micronutrients zinc, manganese, copper, molybdenum, boron, cobalt, chlorine, and bromine are needed in mere traces (ounces per acre or grams per hectare). However, if any of these nutrients is missing or is present only in an unavailable form, plants cannot survive. Also, exhaustion of the essential nutrient in shortest supply will restrict further plant growth, even if other nutrients are abundantly available. But an excess of a nutrient or an imbalance between two or more can also severely harm plants. Furthermore, there are fairly common substances in the environment, both synthetic and natural, that are toxic to plants. And, in addition to substances that act directly on plants, others act indirectly but equally powerfully through their effects on the soil- and water-dwelling microorganisms that are such intimate and influential neighbors of plants.

Water

Although it is not often thought of in such terms, water is the most critical nutrient for plants. It is also the one most often in critically short supply. Water is the major source of hydrogen (required in photosynthesis) and oxygen (generated by photosynthesis and required for respiration), and it plays other critical roles in plants' physiology.

Plants are wet organisms, with internal water contents of 85–95 percent or more as compared to animals' 55–75 percent. Plants take up vast quantities of water through their roots and transport photosynthesized carbohydrates from leaves to other parts in a continuous stream of water within the phloem. Most common agricultural and garden plants consume 300 to 800 pounds (140 to 360 kg) of water per pound (0.45 kg) dry matter produced. Ironically, some desert plants such as Mormon tea (*Ephedra*) and burro weed (*Isocoma tenuisecta*) consume as much as 2400 pounds (1080 kg) of water to produce a single pound of dry plant matter. Under some conditions, this figure can be even higher. On a sunny, warm, windy day, a fast-growing pumpkin plant (*Cucurbita maxima*, *C. pepo*) can consume almost its own fresh weight in water every daylight hour. This is the equivalent of a human drinking about 20 gallons (78 L) of water per hour!

It is crucial to distinguish between water and available water. The ocean contains huge volumes of water, yet due to its dissolved salt content, a person in a life raft could not survive by drinking it. In contrast, fish also require water in their cells to sustain life, yet they survive in the ocean handily. Why? Because the fish have biochemical mechanisms for extracting pure water from the ocean; seawater is *available* to them. Water available to plants includes that found in the soil, liquid water in streams and other water bodies (if free of excessive salts or other toxic substances), and, to a very limited extent, water contained in the atmosphere that can condense onto plants in the form of dew. But plants in the garden often face a situation similar to that encountered by people on a life raft. Even in the desert or during severe droughts, the soil may actually contain much water, but this residual water is so tightly adsorbed by the soil that plants cannot extract it and thus perish of dehydration. (Some soil-dwelling fungi, in contrast, are capable of extracting tightly held soil moisture and may survive in soils too dry for most plants.) Excessive dissolved salts or toxic substances may also make water unavailable to plants.

Although plants have a great need for water, they vary widely in their abilities to extract usable water from their environment and in their abilities to adapt to water shortages or excesses (plate 83). Plants range from extremely drought-tolerant species, such as portulaca, olives, and yucca (*Yucca*), through common yard and garden plants with intermediate water needs, to plants that require high levels of free moisture at all times, such as water lilies (*Nymphaea*) and watercress (*Nasturtium officinale*). A gardener needs to be aware of this variability, as one watering schedule may not be suitable for an entire garden—what is good for one plant may kill another.

Water is capable of dissolving a wide variety of substances, especially electrically charged substances such as salts. Indeed, essentially no water in nature is free of dissolved salt. But salt contents vary widely, ranging from virtually undetectable levels in some spring waters to about 3.5 percent in ocean water to 27 percent in the Great Salt Lake and 32 percent in the Dead Sea. Tolerance by plants of salty water also varies a great deal. Samphire (*Crithmum maritimum*), asparagus (*Asparagus officinalis*), beachgrass (*Ammophila*), and nitrebush (*Nitraria schoberi*) tolerate exposure to significant levels of salt. (A traditional, although not recommended, way of weeding garden areas containing some of these plants is to sprinkle rock salt around them.) Strawberries, carrots, and lima beans (*Phaseolus limensis, P. lunatus*), on the other hand, may be severely injured or killed by even brief exposure to salty water.

Plants use a variety of strategies for survival and growth with little available water. They may expand root systems to extract moisture from a larger volume of soil, or induce biochemical changes in roots to increase vigor of water uptake. Plants may produce tough surface coatings on stems, leaves, and roots to reduce water loss by evaporation. They may slow growth or even enter dormancy to reduce water needs. Plants may accumulate dissolved sugars and organic acids in cells to reduce water losses by evaporation and leakage; this strategy also increases competitiveness of cells for salty or tightly adsorbed water. If a drought is too severe for a plant to adequately adapt, premature flowering and seed production may occur, thus increasing the likelihood of leaving some offspring even if the parent plant perishes.

Gardeners must often concern themselves with these mechanisms. Plants grown indoors or in a sheltered location may be tender—unadapted to moisture stress—and may perish if suddenly transplanted into the sunny, windy garden environment. Even if the soil is moist, unhardened recent transplants may not be able to draw up enough water through their diminutive and often damaged root systems to replace that lost through the leaves (plate 84). Plants started indoors should be hardened off before transplanting, with water gradually withheld and plants progressively exposed to the harsh outdoor environment so as to trigger drought-tolerance adaptations. However, plants should not be overhardened by sudden or severe water deprivation so as to trigger dormancy or premature flowering. Life-long stunting or premature aging are rarely what gardeners want for their plants!

Because one of the major drought adaptations of plants is root system expansion, widely spaced plants generally tolerate drought better than do crowded ones. Furthermore, rapid depletion of soil water by crowded plants can quickly cause water to become depleted during even brief dry spells. The plants most likely to vigorously compete with a plant for water in a garden are other plants of the same species. A very common gardening mistake is inadequate thinning of overcrowded seedlings. Garden plants should not be crowded unless they can be provided a constant and abundant supply of moisture and, possibly, other nutrients.

The physiological effects of *moderate* moisture deprivation can have both positive and negative consequences for the gardener. If water is moderately restricted, fruits such as strawberries, tomatoes, and peaches may be smaller but sweeter and more intensely flavored and colored. Bread wheat (*Triticum aestivum*) grains, beans, and other dry seeds may be shrunken but higher in protein content. Under dry conditions, some ornamentals flower more profusely and the flowers are colored more intensely. Some drought-adapted herbs such as sage (*Salvia*) marjoram (*Origanum majorana*), and thyme (*Thymus*) may be dwarfed but also more flavorful due to higher concentration of aromatic oils. However, few gardeners appreciate hot, pithy radishes; bitter, leathery lettuce; or brown, shriveled zinnias.

Serious water deprivation at flowering time can be especially damaging, causing pollen sterility and severely depressing viable seed production. This effect is especially prominent in corn/maize. With adequate moisture at tasseling time, even drought-stunted plants will form ears with viable seed, although ear size may be significantly reduced. Conversely, even a brief drought at that critical time may result in few or poorly developed ears even on large, robust plants watered well the rest of the season. Prolonged or severe drought causes wilting, leafburn, and defoliation and may irreversibly stunt, or even kill, all but drought-adapted plants. It is striking, though, that many seemingly drought-killed plants, especially deep-rooted perennials, manifest a remarkable ability to revive and continue development when finally given water. These resurrected plants are usually so stunted and deformed, however, as to be undesirable to the gardener.

Some gardeners are horrified by the seemingly brutal practice of pruning perennial plants when transplanting them. In truth, the practice is a kindness to the plants. When dug up or unpotted, plants often lose many of their roots, especially the fine root hairs responsible for most water and nutrient absorption from the soil. If they are transplanted without a corresponding reduction of leaves and stems, water loss by transpiration from the aerial plant parts may exceed the ability of the severely truncated root system to extract soil moisture, even from damp soil. Injury or death from drought stress may then occur. Proper pruning to bring upper plant parts into balance with the reduced roots will help a plant as a whole to recover from transplant trauma.

In gardening books, few instructions are as often repeated as the need for well-drained soil, especially for perennial plants. Excessive moisture can be as devastating to plants as insufficient water—possibly more so. Whereas drought injury requires days or weeks to develop and is often at least partially reversible by supplying water, outright plant death from flooding can be very rapid, within hours or even minutes. A lack of water injures plants by depriving them of an essential substance necessary for their biochemical and physical functioning. But excessive water injures plants indirectly, by depriving them and the soils in which they grow of another critically needed substance, oxygen.

Roots require oxygen just as do leaves and stems. Oxygen is relatively insoluble in water, and the oxygen content of most natural waters is inadequate to support aerobic plant life. With the exception of some aquatics, totally immersing plants in water for a prolonged period will drown them. Even partial immersion, if repeated or prolonged, will generally result in stunted, chlorotic (yellowed), sickly individuals (plate 85). But just as animals vary widely in their ability to hold their breath underwater, so do plants vary in their adaptability to excessive moisture.

Legumes (beans, peas, locust trees, and relatives) and members of the rose family (including roses, apples, raspberries, and peaches) are especially vulnerable to drowning. Rose family plants not only suffer from oxygen deprivation during immersion, but may also release significant amounts of cyanide from their roots, which greatly intensifies the injury. In contrast, many grasses and sedges, including corn/maize, fescues, bluegrass, nutsedge (*Cyperus esculentus*) and its domesticated cousin chufa (*Cyperus esculentus* var. *sativus*), and, of course, rice, are relatively insensitive to flooding and can endure partial or occasional immersion without apparent injury. Even most grasses, however, will show yellowing, stunting, and other effects of oxygen deprivation if subjected to prolonged or repeated immersion. Truly aquatic plants may partially trap and reuse oxygen released during photosynthesis. The excess oxygen is seen as bubbles emerging, as from illuminated aquarium plants. Alternatively, aquatic plants may transport atmospheric oxygen to submerged parts.

Chronic or repeated saturation of soils can also indirectly cause adverse effects on plants through changes in soil chemistry and microbial populations. Anaerobic soils frequently contain toxic levels of substances such as organic and inorganic acids, soluble aluminum and manganese, and hydrogen sulfide. Indeed, the major physiological adaptation of many wetland and aquatic plants to flooding is biochemical tolerance of these otherwise poisonous substances, rather than an ability to withstand oxygen deprivation.

Oxygen

Although a critical nutrient, oxygen can also be very toxic, and all aerobic organisms require biochemical detoxification mechanisms to dispose of excess oxygen. Because plants are exposed not only to atmospheric oxygen (18–21 percent O_2 by volume) but also to much higher local concentrations in their leaves (due to photosynthesis), they must have especially efficient oxygen disposal mechanisms. There also exist especially toxic chemical forms of oxygen that can significantly injure plants at low concentrations. Most common is ozone (O_3), a very reactive and unstable form of oxygen gas, produced by a variety of industrial and natural processes. Ozone harms plants by injuring stomatal guard cells and by damaging membranes of photosynthetic centers, which cause a characteristic flecking or, in extreme cases, metallic silvering of leaves. It also induces stress responses in plants such as synthesis of woody material and toxic self-defense chemicals, much as occurs in response to severe pathogen attacks. In industrial areas or during hot, humid, sunny weather in some urban centers, greenhouse operators install charcoal filters on air intakes to protect their plants from ozone.

Carbon

The main source of carbon for plants is carbon dioxide in the air or dissolved in water (in the forms of bicarbonate and carbonate). Although carbon dioxide comprises only about 0.04 percent of the atmosphere, that figure represents about 850 billion tons (775 billion metric tonnes) of carbon. One would think, therefore, that air-bathed plants would always have an ample supply of carbon. Surprisingly, this is often not the case. In bright light and calm air, some closely spaced, otherwise well-nourished and well-watered plants can so deplete the carbon dioxide close to their leaves that shortage of carbon may actually limit growth. Examples include tropical rainforests;

corn/maize fields on sunny, calm summer days; and crowded, fast-growing, young plants in a greenhouse. Commercial greenhouses may actually provide carbon dioxide fertilization for high-value plants.

Some scientists feel that carbon dioxide–enhanced photosynthesis by plants could help counteract rising carbon dioxide levels in the atmosphere that result from human activities, thus ameliorating the so-called greenhouse effect. So, if plants are to serve as a clean-up crew to remove excessive carbon dioxide from the atmosphere, which species should be planted? You probably don't want to hear this, but the leading candidate seems to be ragweed! Experiments have shown that ragweed (*Ambrosia*) can convert more carbon dioxide into plant tissues per growing season per acre of crop than can any other plant so far investigated (plate 86). More socially acceptable carbon dioxide gulpers are corn/maize, sugar cane (*Saccharum officinarum*), and some other fast-growing tropical grasses.

Symptoms of carbon nutrient deficiency are not well defined, but seem to include retarded growth and development. In fact, it is possible that what we have long accepted as normal plant growth may actually be limited, at least on occasion, by carbon availability. The average gardener can do little, of course, to influence the carbon nutrition of plants. But it might help to avoid overcrowding plants when possible and to maintain a decaying organic mulch layer on the ground to release carbon dioxide on the underside of plants, where it is most likely to become depleted.

Nitrogen

Critical substances in plants, including enzymes and all other proteins, chlorophyll, and vitamins, contain nitrogen. All garden-dwelling fungi, bacteria, protozoans, and animals also require substantial amounts of available nitrogen for survival. Next to water and oxygen, nitrogen is probably the nutrient that most frequently limits plant growth.

There are vast quantities of nitrogen in the atmosphere (approximately 80 percent of air is nitrogen—2000 times the carbon dioxide content) and every plant is bathed in tons of this gas. So why is it so often limiting? Simply put, atmospheric nitrogen is unavailable for use by plants because it is so chemically inert. For plants to use nitrogen, it must first be converted from nitrogen gas (N_2) to nitrate (NO_3^-) or ammonia (NH_3), which requires vast amounts of energy. This conversion process is carried out primarily by industrial fertilizer factories and by nitrogen-fixing bacteria (see chapter 5). Because farms, gardens, and landscaped areas consume most nitrogen fertilizers, industrial nitrogen fixation comprises a major ecological impact of plant raising, albeit indirect. Lightning fixes a much smaller, although occasionally locally significant, amount of atmospheric nitrogen into water-soluble nitrogen oxides that are deposited in rainfall. Temperature also affects nitrogen nutrition. In cold soils, plant roots can readily take up only the nitrate form of nitrogen; ammonia is relatively unavailable. Thus, gardeners and farmers who use ammonia-producing manures, compost, bloodmeal, anhydrous ammonia, or urea as fertilizers may see nitrogen deficiency symptoms on their plants in cold weather, no matter how much they apply.

Nitrogen-deprived plants become tough, stunted, woody, and pale green or yellowish—especially in older tissues, because the plants selectively cannibalize chlorophyll and protein in older leaves and ship the soluble nitrogen compounds thus obtained to younger tissues (plate 87). Carbon-rich carbohydrates, organic acids, and wood fibers also accumulate as the carbon-to-nitrogen ratio of the plant tips too far to the carbon side, causing the plant to dump high-carbon materials that cannot be used for protein synthesis. Under conditions of *moderate* nitrogen deficiency, flowering and fruit production may be enhanced and vegetative growth retarded. Moderate nitrogen deficiency also enhances resistance of many plants to drought, frost, and disease, although the toughening adaptations may make the plants less acceptable as food or ornamentals. Under severe nitrogen deficiency, the whole metabolic machinery of a plant ceases to function effectively.

Excessive nitrogen is a problem rarely encountered in nature but is often found in gardens and lawns cared for by fertilizer-happy homeowners. Nitrogen-overdosed plants become deep green, soft, and they grow very large, but flowering and fruiting are reduced and fruit ripening is delayed. With their watery, succulent, fragile tissues, they also have reduced resistance to drought, frost, pests, and disease. Some plants, especially of the goosefoot family (beets, spinach, chard), accumulate excessive nitrogen as free nitrates, which are toxic and particularly dangerous to young children (causing so-called blue baby syndrome).

Like oxygen, nitrogen also comes in toxic gaseous forms generated at least partially by human activities, with vehicle exhaust and burning of coal being major culprits. Nitrogen oxides are major constituents of acid rain. Peroxyacetyl nitrate (PAN), a sunlight-generated reaction product of hydrocarbons and nitrogen oxides, is a major component of smog, especially in the southwestern United States. This smog causes ozonelike injury to plants, only more severe.

Phosphorus

Plants and other organisms need phosphorus in substantial quantities as an essential component of genetic materials (DNA, RNA), cell membranes, sugars, and carbohydrates (plate 88). Soils generally contain levels of total phosphorus that would be adequate for most plants were it in available chemical forms. However, it often is present in very insoluble compounds—calcium and magnesium phosphate minerals in temperate soils and iron and aluminum phosphates in tropical soils. Thus, available phosphorus is frequently in short supply for plant growth (although less often than is nitrogen, oxygen, or water), especially in humid and seasonally wet-dry tropical soils and in alkaline desert soils.

Phosphorus is relatively immobile and rarely lost altogether from the soil; furthermore, certain soil fungi can mobilize soil phosphates and convert them to forms available for plant use. Plants themselves vary widely in their ability to obtain phosphorus from soil; some may grow quite well on the same ground that causes symptoms of deficiency in others. Temporary phosphorus deficiency may also occur in cold soils, as both roots and soil microbes are then less effective at mobilizing soil phosphorus. Generally, young, fast-growing plants or those grown for their reproductive parts (seeds, flowers, or fruits) require more phosphorus than mature ones or those raised primarily for their roots, stems, or leaves. Fertilizer solutions high in soluble phosphates (starter solutions or bloom boosters) are often used to help overcome transplant shock and to enhance blooming and fruit production.

Although symptoms of phosphorus deficiency are generally not very distinctive, some deficient plants accumulate purple pigments in leaves and stems. The primary damage, however, is quantitative—reduced flowering, aborted fruits, slowed growth, delayed maturity, and poor recovery by young plants from transplant shock. Usually, applying too much phosphorus fertilizer simply results in more insoluble phosphate minerals in the soil or contamination of streams and lakes (see chapter 6), rather than in direct injury to plants. Grossly or repeatedly excessive applications can, however, indirectly harm plants by lowering soil pH (most common phosphate fertilizers are quite acidic). Soluble phosphate reacts with zinc to form a very insoluble compound, and thus excessive amounts can cause zinc deficiencies. Phosphate-containing runoff can also cause problems in lakes and streams by stimulating excessive growth of algae. This can lead to waters becoming anaerobic, cloudy, smelly, and hostile to many desirable organisms such as fish and crustaceans.

Potassium

To maintain the proper composition of internal fluids and activity of many enzymes, plants require potassium. Most temperate-zone soils have adequate supplies of potassium for plants in both available forms (adsorbed by soil particles and by organic matter) and insoluble minerals that can gradually break down into available forms. However, sandy soils and some subtropical and tropical soils may have shortages of both available and reserve potassium.

Visible symptoms of potassium deficiency are rather indistinct but include marginal leaf scorching and spotty leaf chlorosis. However, the primary injury from potassium deficiency is underground: stunted root systems that lead to generally poor uptake of nutrients from the soil, poor growth, and low drought tolerance. Crops that are especially demanding of potassium, such as legumes (clovers, alfalfa [*Medicago sativa*], beans) and root crops (beets, carrots, radishes, turnips), may benefit from supplemental potassium applications even when no overt deficiency symptoms are apparent.

Excessive potassium is a rare problem. Most potassium fertilizers are soluble salts, however, and excessive applications can injure salt-sensitive plants. Potassium can also act antagonistically to magnesium, and overdoses may induce symptoms of magnesium deficiency, even when normally adequate levels of magnesium are present. (Conversely, excessive magnesium can induce symptoms of potassium deficiency.) Also, the chloride part of muriate of potash (potassium chloride), the most common source of potassium in commercial fertilizers, may accumulate to levels toxic to some plants.

Calcium and Magnesium

In soil and in plants, calcium and magnesium are generally associated with each other, both in function and in physical and chemical behavior. Calcium is essential for normal development of plant tissue structure and defenses against disease organisms. Blossom end-rot of tomatoes, celery blackheart, and hollow carrots are disorders caused by calcium deficiency, as are weak, brittle stems of many cutting flowers and watery, soft, decay-prone apples (plate 89). Plants use magnesium for flowering, fruiting, and photosynthesis. It is an essential component of chlorophyll, and deficiency causes a yellowing of leaves that resembles nitrogen deficiency. Magnesium deficiency also causes abortion of flowers and failure to set fruit; this is especially noticeable in tomatoes and peas.

Calcium and magnesium tend to be abundant in alkaline or neutral soils of arid and grassland areas, but may be deficient in acidic woodland soils, especially sandy ones, and very deficient in many tropical soils. Even in soils where these nutrients are abundant, however, an imbalance of calcium and magnesium can be a problem. The two nutrients work against each other to some extent, so an excess of one can cause problems resembling a deficiency of the other. Gardeners may cause calcium-magnesium imbalances by overliming acidic soils in an attempt to raise the pH. (Overliming also can raise soil pH to the point where deficiencies of alkali-insoluble nutrients such as iron, copper, and manganese can arise.) Labels on bags of lime tell whether they are high-magnesium dolomitic limestone or high-calcium limestone. If large amounts of lime are required to correct excessively low soil pH, a gardener should have a soil test done to determine which type of lime (or a blend of both) should be used. To more quickly remedy deficiencies or imbalances of these elements, apply magnesium as Epsom salts (magnesium sulfate hydrate) or calcium as calcium chloride (sometimes sold as a "plant-safe" de-icer for sidewalks and driveways) or calcium nitrate. Be cautious, though; these sources are soluble salts and can cause salt injury if applied to leaves or in excess to the soil. A safer, although slower-acting, source of calcium is gypsum (calcium sulfate hydrate), which will cause neither salt nor pH problems.

Sulfur

In temperate-zone soils sulfur is rarely deficient, and industrial air pollution has made deficiencies even rarer. When sulfur is deficient, though, symptoms generally resemble those of nitrogen deficiency because sulfur deficiency also disrupts protein synthesis. Onion growers in parts of Georgia take advantage of moderate sulfur deficiency to produce sweet Vidalia® onions. The sands in which these onions are grown are very low in sulfur, so the plants can make very little of the sulfur-containing substances that give onions their bite. Thus, even if Vidalia® onions are grown elsewhere, they probably won't be as mild as the real thing exported from Georgia.

Excessive sulfur can be a serious problem, especially in industrialized areas and regions downwind from them. Gaseous sulfur oxides are released by burning fossil fuels, especially high-sulfur coals and fuel oils. Not only is sulfur dioxide directly toxic to plants, severely damaging their photosynthetic machinery, but it also reacts with oxygen and water in the atmosphere to form sulfuric acid, a major component of acid rain. Dilute sulfuric acid itself does not cause serious direct harm to most plants, but it can acidify soil and waters, thus converting insoluble aluminum and heavy metal compounds to more soluble and toxic forms (more on this later). In areas with acid rain problems, gardeners would be well advised to monitor the pH of garden soils and add lime if necessary to adjust it to above 6.2, the lower limit of good growth of most garden plants. However, growers of azaleas, rhododendrons, blueberries, cranberries (*Vaccinium macrocarpon*), and other plants requiring very acidic (pH < 4.5) soils capitalize on sulfur's ability to react with oxygen and water to form sulfuric acid. Adding flowers of sulfur (pure powdered sulfur) or aluminum sulfate to the soil around these plants makes the soil more acidic (plate 90). Most other plants, however, would be sickened by this acid treatment. In anaerobic flooded soils and stagnant waters, sulfur can be converted by microbes into hydrogen sulfide, the infamous rotten egg gas. Hydrogen sulfide is actually more poisonous to aerobic organisms than is cyanide gas, and it is a major cause of the injury suffered by plants in waterlogged soils.

Iron and Related Metals

Iron and chemically similar metallic elements such as copper, zinc, and manganese are required in only trace amounts by plants, yet all are crucial for survival and growth. Although some of these elements are present in large quantities in most soils, they are generally found in the form of very insoluble, and thus unavailable, minerals. The greatest insolubility is at alkaline pH or under warm, highly aerobic conditions, so deficiencies of these nutrients are most common in alkaline soils of deserts and steppes and in highly oxidized (red) tropical and subtropical soils. These nutrients become much more soluble at acid pH and under anaerobic conditions and can be present in toxic amounts in waterlogged acidic soils.

Deficiencies are rare in soils of humid temperate-zone areas, unless they are very sandy or ancient.

Although all of these metals have distinctive and necessary roles in plant metabolism, an excess of one may cause symptoms of deficiency of the others. Symptoms of iron, copper, and manganese deficiencies are roughly similar: stunted plants and yellowed leaves that wither and die prematurely as in nitrogen or sulfur deficiency, except that symptoms appear first and worst on the younger leaves, not the older ones. Zinc deficiency is more distinctive, generally appearing as distorted leaves, death of growing points, and variegated mottling, with red-brown pigmented areas and an uneven chlorosis (plate 91).

Temperate-zone gardeners will rarely observe deficiencies of iron, copper, manganese, or zinc in annual plants. However, perennial fruit and nut trees such as peaches, roses, apricots (*Armeniaca vulgaris*), and pecans seem prone to these deficiencies, especially if raised on sandy or alkaline soils. These deficiencies may be relieved by various nutrient preparations sprayed on leaves or dissolved in irrigation water, or, in the case of zinc deficiency, by driving one or more small galvanized (zinc-coated) nails into tree trunks. Iron, manganese, and copper toxicities are very rare in plants grown in well-drained soils. These conditions, when seen, generally are found in plants grown in acidic, wetland, or some tropical soils; in plants exposed to overdoses of copper-containing pesticides; or in greenhouse-grown plants given excessive doses of fertilizers containing these substances. Overdoses cause general stunting and sickliness of plants and may resemble deficiencies of other metal nutrients, such as an apparent iron deficiency in the case of copper overdose.

Other Micronutrients

Molybdenum and cobalt are required in extremely tiny amounts by most plants, and deficiencies are rarely observed. A major exception is seen in the legume family, plants of which require larger amounts of these elements, apparently on behalf of the nitrogen-fixing bacteria that inhabit their roots. Without adequate molybdenum or cobalt, the bacteria cannot fix nitrogen, and their legume hosts become susceptible to nitrogen deficiency in much the same fashion as other plants.

Boron is an especially interesting nutrient because the window between the required amount and the toxic level is very narrow.

Nevertheless, neither boron deficiencies nor toxicities are commonly observed. When deficiencies do arise, however, the symptoms are dramatic and distinctive. Root crops such as beets and turnips; cole crops such as broccoli, cabbage, and cauliflower (*Brassica oleracea* Botrytis Group); and legumes appear to be the most sensitive to boron deficiency. Symptoms include internal browning or blackening of roots and stems accompanied by softening, hollowing or cracking, and rotting. Legumes may also become stunted, turn yellow, decline, and die prematurely. The specific function of boron in plants is not well known, but it appears to be involved along with calcium in structural development and maintenance. Where needed, boron is generally applied to plants in the form of borax, a naturally occurring mineral—although great care must be taken to avoid overdose.

Two other required micronutrients, chlorine and bromine (usually found in soils as chloride and bromide salts), are very rarely deficient. However, *excessive* chloride is a fairly common problem, often resulting from overapplication of fertilizers containing potassium chloride (muriate of potash), or worse, from contamination by sodium chloride (table or rock salt). Symptoms of chloride intoxication are basically the same as salt damage for most plants, except that some plants undergo changes in flavor or other chemical characteristics before visible symptoms appear. For example, chloride-containing fertilizers cause pronounced chemical changes in tobacco (*Nicotiana tabacum*) leaves, resulting in undesirable changes in the tobacco smoke aroma.

Sodium
The sixth most abundant element in the Earth's crust is sodium. A vast quantity of sodium is dissolved in seawater, and it is found in many natural minerals. It is not a nutrient required by terrestrial plants, however, and sodium can be harmful if present in other than very low concentrations. Harmful levels of sodium usually reach lawns and gardens via runoff from rock salt de-icers used on sidewalks, driveways, and streets; by way of salt spray from the sea; or in irrigation water that has passed through salt-loaded water softeners (plate 92). In arid areas, some irrigation water may contain high levels of sodium leached from natural mineral deposits.

Sodium is directly toxic to most plants, acting as an antagonist to potassium, and, even worse, it devastates soil structure. The effect of sodium on soil structure can be more severe than compaction, and it can puddle soil to a dense, airless, rock-hard or gummy mass, useful for little besides making bricks and pottery. Sodium destroys the ability of soil particles to repel each other and remain fluffed up despite the tremendous pressure of the soil's own weight. The medical saying: "Hard water, soft arteries; soft water, hard arteries" can be applied to the garden as well, by replacing "arteries" with "soil." Perhaps gardeners should get rid of their water softeners—or at least install bypasses for drinking and plant-safe irrigation water. Sodium chloride de-icers can be replaced with more expensive potassium chloride or, better yet, calcium chloride, which might be cheaper in the long run because they do much less harm to plants and soil and cause less deterioration of concrete structures. Oceanside gardeners can hardly stop salt spray, however, and may have to content themselves with growing sodium-tolerant plants or adding gypsum to their soils, which helps counteract the soil-degrading effect of sodium.

Other Mineral Substances

In addition to taking up substances having nutritional properties, plants are capable of absorbing and accumulating almost any substance in the soil-water solution. Some of these substances seem to be innocuous: silver, nickel, strontium, and gold, for example. In fact, there is a whole field of bioprospecting for metal ore deposits, more popular in the nineteenth and early twentieth centuries than today, that relies on indicator plant species and chemical analysis of plant tissues to detect mineral deposits in the plants' root zones.

Other materials are not so harmless. Especially in strongly acidic or waterlogged soils, there can be dangerous levels of soluble aluminum, fluoride, and toxic metals such as selenium, arsenic, chromium, lead, mercury, and cadmium. These substances can harm many plants themselves, but some plants appear to be immune to the toxic effects of these substances and may selectively accumulate them in high concentrations, thus poisoning animals and humans who eat them. Several of these species are being considered as potential allies in decontaminating toxic waste sites. For example, the Chinese brake fern (*Pteris vittata*) is a very selective but avid scavenger of arsenic, concentrating that element in its tissues by up to 130 times the level in soil. Needless to say, plants that cause stomachaches, or worse, in herbivores will generally be avoided by them. Thus, plants that accumulate toxic substances may have a selective advantage over more palatable neighbors.

Plant Communities

On the salt flats of Utah, along the backbones of mountain ridges, and on sand dunes and beaches the physical environment stands out, but in most familiar habitats the vegetation predominates. Plants form towering forest canopies or continuous seas of prairie grass, and together with other organisms in the habitat they form communities. Not all species in a community come into direct contact with each other, but there is the potential for direct or indirect interaction, and each species affects its habitat and community in some way.

Because plants are the most conspicuous organisms in the great majority of communities, habitats are often described in terms of the currently or historically dominant vegetation. For example, many dry upland sites in the American Southwest are classified as juniper-pine woodlands. Farther north, a swath of boreal forest dominated by spruce (*Picea*), fir (*Abies*), and pine (*Pinus*) sweeps across Canada and into Alaska. In the United Kingdom, heathland vegetation dominated by dwarf shrubs such as heather (*Calluna vulgaris*) and bilberry (*Vaccinium myrtillus*) grows widely, particularly in Scotland and in uplands farther south. Much of the American Midwest is labeled as tallgrass prairie on vegetation maps, despite the fact that most of the original prairie (more than 98 percent) has been replaced with human-altered landscapes. Similar replacement of native vegetation with human-altered landscape has taken place in the Atlantic coastal forests of Brazil and the plains of northwest Europe.

Gardens, too, are communities whose nature varies dramatically depending on geographic location, microclimate, soil, and the many and diverse species that are present (plate 93). Organisms find their way into gardens in any number of ways. Some are added by the gardener (intentionally or otherwise); others are transported by air or water after the garden is planted. Regardless of when and how they arrive, the organisms in a garden form a community of interconnected populations whose whole is far greater—and far more interesting—than the sum of its parts.

Community Structure

No two communities are identical in their physical and biological characteristics. Within a single garden, for instance, conditions may vary considerably over even short distances. Furthermore, over time conditions can vary in sometimes unpredictable ways as water and nutrient concentrations change and organisms come and go.

Variation in space. Although a meadow, a forest fragment, a creekside community, and a shrub thicket might exist within a short distance of each other, physical characteristics across this meadow-to-thicket transition are likely to vary substantially. There will be differences in soil water content and nutrient availability, light levels at the ground surface, and susceptibility to erosion, among other factors. Furthermore, these habitats likely support different combinations of plants, fungi, vertebrates, and so on, which then accentuate these habitat differences. Physical and biological variation within a typical garden may be less extreme and less obvious than in the natural world, but the variation is important nonetheless. For example, in a garden planted at the edge of a deciduous woodland, there is likely to be a light-intensity gradient at right angles to the forest edge. Soil acidity and organic content may also change across this transition as a result of differences in leaf litter.

When a community is viewed from the ground, vertical stratification, or layering, is often evident. In a New England deciduous forest, for instance, the canopy may include sugar maple (*Acer saccharum*), American beech (*Fagus grandifolia*), eastern hemlock (*Tsuga canadensis*), and eastern white pine (*Pinus strobus*), which vary in height, diameter, and branching patterns. Below the canopy are smaller individuals of the canopy species as well as birch (*Betula*) and other tree species. Shrubs, saplings, forbs, grasses, mosses, and other plants form a discontinuous layer at and above ground level. To some degree, all communities are stratified, but distinct layers are particularly well developed in tropical rainforests, where up to five vegetation layers often develop. Gardens are inevitably stratified, if only because individuals of the same species do not all grow to the same height, nor do they have identical architecture of stems and branches. More importantly, plantings of different species, a scheme found in nearly all gardens, results in stratification that is largely under the control of the gardener, as when a low-growing border is planted in front of midheight perennials and taller background shrubs.

Animal species take advantage of this stratification. Deer are restricted to a narrow zone at ground level; squirrels, on the other hand, move freely throughout the entire vertical structure. Other species may subdivide the habitat more finely; for example, different butterfly species tend to favor garden plants of particular heights. All other things being equal, a community with several layers can support a greater diversity of species than can a community with only one or two layers.

Communities vary horizontally as well as vertically (plate 94). Horizontal variation exists because of a combination of biological and physical characteristics and is best viewed by walking through a community. Uneven topography creates depressions that are generally moister and shadier. Fallen logs add humus and form sheltered sites. Slopes have thinner soil and receive either more or less solar radiation, depending on which compass direction they face. Plants of different species, ages, and sizes grow in different groupings. Herbivores and pathogens may cause a patchy distribution of healthy and unhealthy plants. Localized disturbances, such as fire or digging by rodents, may cause small patches that differ from surrounding areas.

Plant communities change from one to another, as in the transition from meadow, forest fragment, creekside, to shrub above. A change from meadow to forest might be abrupt if the edge is maintained artificially or if soil type changes suddenly, or adjacent communities may blend together subtly. Transition areas where two communities intermingle are called ecotones (plate 95). These areas tend to have some characteristics in common with the communities from which they are formed and some characteristics that are unique. An ecotone may have a microclimate that differs slightly from those of adjacent communities, or it may include unique plant or animal species. A garden edge created by a shovel or a rototiller is usually a rather sharp ecotone formed between the garden and the habitat in which it is set, at least in the beginning. But over time, as plants spill into and out of the garden, as burrowing animals mix soil along this edge, and as other changes take place, the ecotone may become less distinct. In addition, not all new gardens are cut cleanly out of existing landscapes. An aquatic garden planted along a pond edge, a rock garden put in on a rocky slope, and woodland flowers planted at a forest edge are all likely to have indistinct boundaries with adjacent communities.

To some extent, spatial variation in gardens is under the control of the gardener and tends to be more limited than in natural communities. However, even slight differences in topography, soil texture, soil chemistry, shading, and countless other factors can result in significant variation across a single garden plot. These differences may be purposely created by a gardener or they may be unintentional. One example of unintentional variability is seen in gardens that are planted around house foundations. Even if such gardens are level, uniformly shaded, and have been tilled, fertilized, and mulched in essentially the same manner, conditions below ground can be surprisingly variable. This may result from how much of the original topsoil was removed (often all or most) and then replaced (often little or none); from the amount and type of fill that was added; and from the amount of gravel, chunks of concrete, pieces of roofing material, and discarded nails and pieces of glass that were purposely or accidentally buried. These and other kinds of debris can alter soil's texture, chemistry, and ability to hold water and to nurture microorganisms, invertebrates, and plants. Other factors out of the control of gardeners may include differential shading by existing trees and buildings and past use of different areas. In Steve Salt's lawn, for example, during drought conditions two parallel strips of poor plant growth are observable, the result of soil compaction and gravel from a road not used in several decades.

Habitat variation can cause problems for growers who wish to raise many plants of a single species. This is especially true when planting garden plots or fields that were originally heterogeneous (such as when roses, which are sensitive to wet soil, are planted in an area having uneven soil drainage) or when growing plants that have narrow requirements. However, gardeners can also take advantage of variation, because different kinds of plants grow best under different conditions. In the Northern Hemisphere, gardens planted on south-facing exposures or around building foundations (the above-mentioned problems notwithstanding) may allow warm-climate species to grow farther north as a result of increased solar insolation or protection. However, some warm-climate perennials grow better on the north side of buildings because of less-dramatic temperature fluctuations in winter.

Gardeners frequently increase habitat diversity by creating high, dry areas or low, moist areas; increasing shade with trellises or tall plants; and adding sand, fertilizer, or other amendments to parts of the garden to vary soil texture, pH, or fertility. Think of gardeners who grow azaleas and acidify the soil immediately surrounding these shrubs or those who plant potatoes in small hills they have created. Increased variation may lead to unexpected benefits. Growing plants with different blooming times, flower morphology, or color may also attract a greater diversity of pollinating insects and birds.

Variation in time. Community structure changes over time as well, with such changes typically following disturbances. These range in extent and intensity from a footstep in a garden to a spring hailstorm to a sudden mudslide. The biological effects of disturbance are varied and sometimes surprising. A foot trail through a backyard woodland may trample plants, compact soil, and erode the surface so severely as to eliminate the organic layer and prevent revegetation for decades. In contrast, a hurricane roaring across an Everglades wetland dominated by sawgrass (*Cladium jamaicense*) or a tornado across a Midwestern oak woodland may do relatively little long-term damage, although the area may appear devastated immediately following the storm.

Regardless of its intensity or duration, disturbance is sometimes detrimental and sometimes beneficial to plants and animals. In jack pine (*Pinus banksiana*) forests in northern Canada, for instance, Steve Carroll found that the average frequency of fires in any particular area was about once every forty-five years. Cooler surface fires burn off the lichen layer and expose the mineral soil, but do little damage to the pines. Hotter fires burn the lichen layer and part or all of the pine canopy as well. Thus, the immediate effects of fire are detrimental to the lichen population, whereas fire has a dramatic, positive effect on jack pine reproduction. When cones are closed, they are rock-hard and the seeds cannot be dispersed; but the cones are opened by intense heat. Therefore, hot canopy fires open cones, cause large numbers of seeds to become available, create a suitable mineral seed bed, and increase the amount of light reaching the ground by removing part or all of the canopy (plate 96). Regrowth starts quickly, and unless otherwise disturbed, a jack pine–lichen community is re-established.

When vegetation is disturbed, whether by natural phenomena or human activity, it may take tens, hundreds, or even thousands of years for something like the original plant community to be re-established. This process is called succession, and the ultimate community that is expected to develop (barring further disturbance) is a climax community. Depending on habitat, particular species predominate in a climax community, and as individuals of these dominant species die they are replaced by other individuals of the same or other dominant species. The classical concept of succession also incorporates predictable stages that follow one another, each having characteristic plant and animal species.

Succession in action can be observed on abandoned agricultural land. In much of the northern and eastern United States, in the first years following abandonment annual weeds become established, in many cases very quickly. But the changes in the physical environment that favored these annuals—in particular, the sudden availability of habitat with little or no competition—are not permanent. As early colonizing species become established, they alter the environment: invertebrates burrow through, fertilize, and aerate the soil; birds nest in or disturb the ground vegetation as they search for seeds or insects; and pioneer plants create shade, block the wind, and enrich the soil. These altered conditions may then favor new species—to the disadvantage of the pioneers. Perennial plants then gain a foothold, eventually outcompeting the annuals established earlier. The shift in successional stages is not wholesale replacement, however; pioneer species disappear plant by plant and species by species, not en masse. Similarly, later-establishing species do not arrive in great waves of colonists. Given enough time, herbaceous perennials tend to gain the upper hand—that is, until woody species become established.

But, as with so many other natural processes, things are not always so straightforward. As ecologists have come to better understand the complexities of organisms and their habitats, the idea of a predictable climax community has been modified. Succession is now viewed as a dynamic process in which roughly comparable endpoints are reached given similar circumstances (for example, similar habitat and climate). Details of the changes and of the plant and animal species that predominate throughout this process are now thought to vary as a result of sometimes minor differences in habitat, time and severity of disturbance, weather patterns in the years following the disturbance, and other factors. Because different species do not respond to these factors in the same manner, details of succession can vary even among sites in close proximity to each other.

Succession does not always proceed as far as it could theoretically, but may be set back, especially by repeated disturbance. In a forest, if trees at intermediate successional stages are cut, thereby opening up the habitat and increasing light levels at the ground, pioneer species may re-establish and reset the successional clock. As a result, the development of a mature forest will be delayed. In fire-prone habitats, such as California chaparral, the successional clock may be repeatedly reset, allowing pioneer and early-establishing species to periodically regain a foothold.

Human intervention can also cause succession to proceed further than it would under natural conditions. For example, this occurs when we extinguish lightning-caused fires in fire-adapted plant communities. In cases such as this, plant communities that were formerly less common become widespread. This has happened in much of the American Midwest. By extinguishing lightning-caused fires, European settlers allowed closed oak-hickory forest to establish on sites that had formerly been dominated by fire-supported savanna. (Although Americans also cut much forest to make land available for agriculture and, later, cities, today there is relatively more closed forest and less open woodland than when European settlers arrived.) Development of irrigation projects in arid regions may permit establishment of mesophytic vegetation and successional trends that would not otherwise occur. Removal of large herbivores has also altered succession; for example, reduction of elephant herds in eastern Africa has permitted thorn forest to invade large areas formerly dominated by savanna.

Gardeners who get behind in their weeding quickly discover that gardens are not climax communities. Even well-tended gardens are subjected to constant pressure from invasive species—those plants able to establish quickly and compete with existing plants, especially on disturbed sites. Abandoned gardens or unplanted areas are usually colonized by opportunistic species—"weeds," to most of us. During the first few years following abandonment, cultivated garden annuals generally disappear. Some of the planted perennials may remain, but eventually, these too may disappear, although some may persist, reproduce, and become naturalized. Over time, the garden plot will likely support an increasing number of locally aggressive native and introduced species.

Although disturbed or abandoned land may seem over time to return to its natural state, the fact is that the community is rarely restored to its original condition. Well-adapted introduced species, once established, usually remain part of the landscape. If the land has been severely altered, native vegetation may not re-establish at all—at least not without assistance from humans. Those interested in re-establishing natural communities are part of the growing movement of restoration ecology. Restoration ecologists sometimes attempt to reintroduce single species to areas where these were once important, as in the recent reintroduction of wolves to the Yellowstone ecosystem (plate 97). Restorationists have also become increasingly interested in re-establishing whole habitats, using management tools such as fire, physical removal of introduced species, and planting of native species. This branch of ecology is still in its infancy, but interest in small- and large-scale projects is increasing.

Conditions vary not only over the time span of decades and years, but across the seasons. As the soil warms in the spring, early wildflowers such as spring beauty (*Claytonia virginica*) and Dutchman's breeches (*Dicentra cucullaria*) emerge through last year's leaf litter. Occasional insects buzz overhead, the first spring peepers (*Pseudacris crucifer*) are calling, and red-winged blackbirds (*Agelaius phoeniceus*), back from their winter grounds, are singing from a nearby pond. As the canopy closes and less light reaches the ground, the late spring forest floor becomes dimmer and less colorful as spring wildflowers switch from flower to fruit production. By now the birds' nesting season is well along, and an occasional broken eggshell can be seen at the base of an oak or hickory—whether from predation or a successfully hatched egg it is difficult to say. As summer draws to a close, the spring ephemerals are nowhere to be seen, their leaves having died back. Many birds are active, but they seem less driven as they search the ground for seeds and tree trunks for insects. Throughout this time and into the autumn, the plants, fungi, microorganisms, birds, mammals, and other organisms that together form this community have been interacting in countless ways as each attempts to obtain needed resources. In the natural world, the timing of these seasonal activities may vary slightly or dramatically from one year to the next, depending on factors such as the weather.

To a limited degree, gardeners can control timing of developmental events (phenology), especially by planting earlier or later in the year. Care must be taken, however, because plants that are put in too early in the growing season may die from frost or desiccation, whereas those planted too late may have insufficient time to mature or flower or may be devastated by large insect populations late in the season. Correct timing is especially critical to farmers and nursery operators, because a mistake can result in significant economic loss.

One way for home gardeners or small-scale growers to manipulate phenology is by starting plants indoors, in a greenhouse, or in a cold frame.

By giving plants a head start under milder conditions, the growing season can effectively be lengthened. Gardeners can also extend the season by covering plants with insulating blankets in autumn or by intercropping species that have different maturation times. For example, during the hot summer, young strawberry plants benefit from growing among okra, a source of shade that facilitates their establishment (plate 98). Manipulating phenology can have other benefits as well. For example, losses to herbivores in annuals can be minimized by altering planting times to avoid insects that tend to appear at particular times each year.

Biodiversity

As commonly used, the term *biodiversity* is the number of species found in a community or habitat. When the term is used in this way, rare species and abundant species are counted equally. Ecologists, however, often distinguish between species richness—the total number of species—and species diversity, which incorporates the total number of species as well as the relative abundance of each. For simplicity, in this book we use these terms interchangeably.

Species diversity may be relatively low in harsh environments, such as salt flats. Diversity may be astoundingly high in tropical rainforests, where an area of a few acres (a hectare) can include several hundred tree species alone. Although tropical rainforests occupy less than 10 percent of the Earth's land area, they are thought to include approximately 50 percent of its species. Even in a human-affected landscape, species diversity can be surprisingly high. For instance, a class of Steve Salt's biology students found more than a hundred plant species in a 100 foot × 100 foot (30 m × 30 m) section of periodically mowed lawn.

The biodiversity of the Earth is truly awe inspiring. Estimates of the total number of species on Earth range from a few million to as many as 100 million, with most estimates in the vicinity of 10 million. To date, fewer than 2 million species have been named. Mammals and birds are reasonably well documented, whereas scientists believe that only a tiny fraction of bacteria, soil invertebrates, and fungi are known. Concerted efforts are now underway to collect and identify Earth's biota.

Like community structure, species diversity varies through space and time. Communities having more heterogeneous habitats and more complex vertical stratification generally support greater numbers of species. This is true, in part, because a greater number of unique micro-habitats are available for species to occupy and because more species having narrow requirements can be packed into a small area. One factor that affects diversity is the total amount of carbohydrates that a habitat's plants can produce. Severe or barren habitats such as alpine boulder fields and mine tailings high in heavy metals can support little vegetation. This reduced plant biomass, in turn, generally supports relatively fewer organisms of other types.

Most gardeners need not face the challenges posed by extreme environments, but neither are all gardens characterized by dark, fertile, well-drained loamy soils. Gardens planted on sites that formerly were coniferous forest are likely to have acidic soil that is leached of nutrients. Soils in glaciated terrain are often coarse and rocky. Gardens in dry climates may lie above salt pans caused by high rates of evaporation. Gardeners can sometimes overcome these and other challenges by altering pH, fertilizing, turning in organic matter, and various other strategies, but not all problems are easily solved.

Biodiversity is affected by the size of an area, as well as by biological structure and abiotic site conditions. All else being equal, if two regions have essentially the same habitat, the one that is larger is expected to support greater diversity. Biologists first formalized the relationship between area and diversity for oceanic islands and then extended it to other types of "islands" (relatively small, isolated habitats surrounded by larger, markedly different habitats) such as mountaintops and lakes. An island's diversity is affected not only by its size, but also by its distance from other islands and from the "mainland" (a relatively large area at a distance from the island that supports some of the same species as the island and whose habitat often shares similarities with that of the island). For example, large islands close to the mainland have greater diversity than large islands farther from the mainland. This is mostly due to the fact that more immigrants from the mainland are likely to arrive successfully on a nearby island than on a distant island. If a large forest tract is viewed as a mainland (and therefore as a source of species) and an isolated forest fragment is viewed as an island, the biodiversity of that fragment—whether of plants, birds, or invertebrates—will be affected by its size and distance from the mainland forest. This understanding is now central to decisions about habitat preservation.

Gardens, too, can be thought of as islands. Two gardens planted in adjacent yards, perhaps even on opposite sides of a fence, may share pools of pollinators and herbivores by virtue of their proximity (that is, their inter-island distance). Gardens that are farther apart, on the other hand, or that are farther from a wildflower meadow (the mainland) may attract fewer of these visitors. Similarly, a garden planted adjacent to a forest edge will attract more woodland residents than a garden planted in the middle of a large lawn (plate 99). With this in mind, gardeners can have at least some effect on the types and numbers of insects, songbirds, weeds, and other organisms that find their way into their gardens based on where gardens are planted in relation to nearby fields, woodlands, other gardens, and so on.

An area's diversity (both its number of organisms and types of species) is not fixed but fluctuates as species are gained and lost. Local extinction often follows habitat deterioration. For instance, aquatic organisms that require cold, well-oxygenated water are likely to be eliminated from lakes that receive large volumes of warm wastewater high in dissolved nitrogen and phosphorus. Such sudden inputs of nutrients often result in large populations of oxygen-demanding microorganisms, such as algae, which alter the environment to the detriment of its original inhabitants. A habitat dominated by algae, diatoms, and anaerobic bacteria, but poor in fish and shorebirds, might then exist.

Another reason for species loss is the edge effect, which relates the amount of perimeter a site has in relation to its area (a function of the fragment's shape; plate 100). For example, consider two 100 square foot (9.3 m²) gardens. A square garden that is 10 feet (3 m) on a side would have 40 feet (12 m) of edge, whereas a 5 foot × 20 foot (1.5 m × 6 m) rectangular garden would have 50 feet (15 m) of edge. One reason why edge effects matter has to do with the behavior of organisms in which the site is situated. For example, if predators of ground-nesting birds are unwilling to penetrate more than a short distance into a fragment, birds nesting near the edge of the fragment are at greater risk than those nesting nearer the center. The implications of shape are relevant to gardens as well. For example, small herbivores that feed in the garden but do not live there will have easier access to plants in a narrow garden. Similarly, weeds that spread by runners may be better able to grow into a narrow garden and to compete with plantings throughout the garden. Garden plants growing along the edge of a plot in an open lawn are also more susceptible to dehydrating winds; on the other hand, they may receive more sunlight due to decreased competition with other garden plants.

Interdependence among the many species in a *diverse* community has sometimes been cited as a source of stability in communities, but not all biologists agree on this point. One challenge to this idea is that species in a community are not equally important, and thus the loss of one species might have a more profound effect on community structure and stability than the loss of another species. Those species whose activities or ecological roles have the greatest effect on community structure and diversity are called keystone species. Bison in pre-European North American grasslands are considered keystone species, as are krill (*Euphausia superba*), marine crustaceans that are abundant in Antarctic waters and that are fed on by penguins, seals, fish, seabirds, and other organisms. In beds of Pacific kelp (*Macrocystis*), sea otters (*Enhydra lutris*) act as keystone predators by regulating populations of sea urchins (*Strongylocentrotus purpuratus*) that otherwise would overgraze the kelp. Loss of keystone species generally has profound effects on communities, for example, releasing one or more of their prey species from predation pressure, thus allowing the prey populations to increase. One outcome of increased population size of one or a few species is a further decrease in diversity due to, for instance, increased competition for resources.

The importance of diversity in gardens is only partially understood, in part because gardeners work hard to prevent unplanned increases or decreases of garden species. One benefit of planting many species, however, is the likelihood that doing so will minimize the buildup of herbivores and parasites of any particular species. In the next chapter, we look at interactions among organisms, including herbivory and parasitism, in greater detail.

5

Interactions Among Garden Organisms

Garden organisms interact with each other in a seemingly endless number of obvious and not-so-obvious ways. The better a gardener understands these interactions, the wiser the decisions that he or she can make about plant choices, cultivation, and pest control. Two plants growing side by side may compete with each other for limited soil nutrients, pollinators, or sunlight. But the shade they cast may also provide protection from excessive light and heat. Elsewhere in the garden, one plant may provide physical support for another, while benefiting from a third plant that acts as a magnet for leaf-eating insects.

A common and convenient way of categorizing interactions among species is based on whether the interacting organisms have net positive or net negative effects on each other. When a caterpillar defoliates a garden plant, the larva receives a benefit and the plant is damaged. Herbivory, therefore, can be viewed as a win-lose relationship, as can classical predation and parasitism. All three are variations on the theme of exploitation, with herbivores, predators, and parasites in the role of consumers (plate 101).

Similarities among these relationships may be clear but their differences are less so. Classical predators are defined as those that capture, kill, and consume their prey. Parasites, in contrast, remove small amounts of tissue from living hosts, in some cases doing so over a substantial portion of the host's life. Parasites generally do not kill their hosts outright, but instead may cause them to succumb more quickly to other causes of death or decrease their reproductive success. If predation and parasitism represent two endpoints on a kill-to-weaken gradient, how can organisms that feed on plants be classified?

Birds that consume seeds are classified as herbivores. But they kill those seeds as surely as do swallows that swoop down on flying insects. Eating seeds, then, is ecologically equivalent to predation. Phloem-tapping aphids are also classified as herbivores. However, aphids can just as easily be thought of as parasites because they feed on small amounts of plant tissue and do not kill plants outright but—especially if present in large numbers—can weaken their hosts sufficiently to lead to an early demise. If the parasite label seems unfounded in this case, compare the plant-aphid relationship to that between a human and a tick.

Competition is viewed as a lose-lose relationship, as interacting organisms each pay a price. Even in cases in which one competing individual wins, for example by obtaining an indivisible resource such as a morsel of food, the winner had to expend energy and take risks that would have been unnecessary had there been adequate food for all. Mutualism, in which both parties benefit, can be characterized as win-win. Although the win-lose paradigm is helpful, categorizing interactions too rigidly oversimplifies the complexities and subtleties of the natural world. In truth, interactions among organisms are fluid and change through space and time. Foraging birds may consume nearly all of a sunflower head's yield of seeds, but the same birds also may have eaten large numbers of weed seeds earlier in the growing season, to the benefit of garden plants—and the gardener—if not the weeds. Carnivorous garden insects may consume many garden pests, but they may also eat large numbers of spiders and beneficial insects.

Yet another limitation of viewing the world through win-lose glasses lies in the false idea that species interact only in pairs. Life in a prairie pothole, a Canadian bog, or a Sacramento Valley garden is not a one-on-one wrestling match. Reducing interaction dynamics to species-species pairs greatly oversimplifies the natural world. An insect feeding on a garden plant is simultaneously hunted by its predators and weakened by its parasites; it competes with other herbivores for choice food plants; it is hindered in its feeding by chemical and physical defenses in the plants it feeds on; and it competes with other members of its species for the best mates and locations for egg-laying. Even a relatively small garden is a miniature ecosystem that includes a surprising diversity of organisms. Some of these organisms—including, we hope, our garden plants—are permanent residents, whereas others may come and go in search of a meal or a mate.

Species interact with each other in many ways, and the more completely we understand these interactions, the less likely we are to unwittingly harm beneficial organisms or provide aid to those we wish had never arrived in the first place.

Competition

A lack of sufficient resources is central to competition. Free-ranging herbivores can compete over vast areas as they seek water holes or forage. Migrating birds traverse thousands of miles, competing for the best sites on nesting grounds. For organisms such as plants or corals, which are permanently rooted in or attached to their substrates, the arena in which competition occurs may be quite restricted. For instance, two garden plants may be growing in soil that is generally rich in nitrogen but that has pockets with lower nitrogen levels. If both plants' roots enter the same nitrogen-poor rooting space, competition for nitrogen can occur in spite of its general abundance in the soil. Gardeners can reduce competition by spacing garden plants appropriately, uniformly providing adequate resources such as water and fertilizer, and weeding out competitors.

Ecologists divide competition into intraspecific competition, in which individuals compete against others of their own species for resources and mates, and interspecific competition, in which individuals compete against members of other species. Differences in the intensities and effects of these two kinds of competition can be significant, and both are important in the garden.

Competition can also be defined in terms of *how* it occurs. In exploitation competition, individuals deplete a resource that is in limited supply, making it less available to others. This type of competition does not involve confrontation or direct challenge. If one butterfly removes the nectar from the open flowers in a garden or meadow, a butterfly that arrives later may find inadequate nectar for its needs. Plants commonly engage in exploitation competition over soil nutrients, water, and light.

In interference, or contest, competition, individuals limit or prevent others from using a resource through direct and sometimes aggressive confrontation. Dramatic examples of interference competition are common in the animal kingdom in contests over mates, food, and territories. But interference competition does not necessarily require bloody battle; many encounters among animals involve ritualistic displays that end short of physical combat, as when birds define their territories using displays or song.

Plants sometimes engage in a form of chemical warfare in which one plant releases a chemical that has harmful effects on other nearby plants. This form of competition, known as allelopathy, is generally interpreted as a form of interference competition (plate 102). Black walnut (*Juglans nigra*) is perhaps the best known allelopathic plant. It produces juglone, a chemical that inhibits and even kills plants such as eastern white pine, paper birch (*Betula papyrifera*), alder, many wildflowers, and garden plants such as tomatoes and alfalfa. Although hostas, bleeding hearts (*Dicentra*), daylilies (*Hemerocallis*), and many grasses and bulbs will grow successfully beneath black walnuts, homeowners who have a choice would be wise to place their gardens at a distance from these beautiful trees. Other allelopathic plants include Bermuda grass, which produces chemicals in its roots that suppress competitors; lamb's-quarters, which releases oxalic acid as it begins to flower; quackgrass; and Canada thistle (*Cirsium arvense*). Although allelopathic plants do not square off and assault each other, the central premise of interference competition—the direct suppression of others—is clearly at work.

What Makes a Plant a Strong Competitor?

No matter the garden or the gardener, unwanted plants (weeds) will appear sooner or later. However, not all uninvited plants grow equally vigorously once established. An experienced gardener who spots an errant milkweed in a vegetable garden may pull it out but probably won't panic. But one who sees quackgrass growing across the lawn toward the garden plot or field bindweed twining its way up a favored perennial is likely to drop everything else and attack these noxious weeds immediately (plate 103). Several years ago, Steve Carroll planned a field experiment using bur cucumber (*Echinocystis lobata*), a troublesome weed in the eastern United States. When he asked permission to plant this climbing plant adjacent to a university-owned corn field, the site manager became agitated. Permission was granted—grudgingly—but only after a promise was made to bag individual fruits before they opened to prevent even a single seed from dropping to the soil surface.

What is it that makes one plant a mild competitor and a relatively inconsequential garden problem and another plant a scourge? And what makes a weed a weed in the first place? Although there is no single set of characteristics that describes all troublesome weeds, a plant that has more than a few of the traits described below is likely to compete vigorously and therefore be a problem in the garden.

Ability to colonize disturbed ground. Although few gardeners would describe their carefully tended plots as disturbed ground, from an ecological point of view, that's exactly what gardens are. In the spring, just when many seeds are physiologically ready to germinate, gardeners go out of their way to break up and turn over the soil surface. Seeds that can take advantage of such conditions are likely to do well. Working the soil also brings buried seeds to the surface, where many have an increased probability of germinating.

As mentioned in the discussion of succession in chapter 4, the colonization of disturbed ground is also important in nature. When a tree falls over—lifting and exposing soil—or when ground is denuded of its vegetation by flood, fire, animal activity, or other disturbance, an opportunity is created for plants (pioneers) that can quickly take advantage of this unoccupied space and increased access to light. Many weeds are pioneer species. Although weeds are often thought of in the context of the garden, look closely along roadsides or on bulldozed piles of soil at construction sites. Growing there will likely be some of the same species that invade gardens, such as pokeweed (*Phytolacca americana*), pigweed (*Amaranthus retroflexus* and relatives), ragweed, and smartweed (*Polygonum*). British gardeners similarly find garden weeds in nearby disturbed sites, weeds such as dandelions and lamb's-quarters. Other localities will have their own cast of opportunistic weeds, but their characteristics will be similar to those listed above. Over time, these colonizers modify the environment by adding organic matter and creating shade, ultimately making it more suitable for other species, and, ironically, less suitable for themselves.

Early growth. Many weeds take advantage of newly available garden space by germinating or putting on new growth early in the season, perhaps before the gardener has found time to begin work in earnest. Weeds may accomplish early growth in a variety of ways. For example, many weed seeds such as annual grasses can germinate in the cold soils of early spring.

Other annuals, including black medic (*Medicago lupulina*) and chickweed, germinate in autumn, establish a rosette, go dormant, and are ready for vegetative growth when the spring warmth arrives.

We demonstrated the importance of timing in a simple greenhouse experiment involving annual rye grass. In this experiment, plant density was kept constant, with the plant of interest (focal plant) in the center of the pot and five competitors planted around it. Timing was manipulated by planting the focal plant either four days before, on the same day as, or four days after the competitors. The plants were then allowed to grow under greenhouse conditions, and differences among the three treatments were compared after two months. In the three early pots (focal seeds planted first), two of the three focal plants had flowered by the end of the experiment, compared to only one of the three simultaneous focal plants and none of the late focal plants. The time of planting also affected the final size of the focal plants, as focal plants in the late pots were only 30 percent as heavy as focal plants in the early pots—yet another example of competition at work. The results of this experiment raise an interesting question. Because time is not a resource taken up and used by plants, which factor or factors in the pots actually caused the size differences that were observed? In many cases, even when competition is clearly at work, the resource over which competition is occurring may never be known.

Rapid growth. Plants that germinate early but grow slowly might only cause minor problems in the garden. However, many weeds grow quickly— they grow like weeds, in fact. As a result, they are able to usurp soil resources or block incoming sunlight to the detriment of plants attempt- ing to grow beneath them. Early vegetative growth also often leads to early flowering and seed production, a combination that makes weeds even bet- ter competitors.

Some weeds take advantage of their rapid growth by growing on, up, and over already established vegetation. In the southern United States, kudzu (*Pueraria lobata*), a purposely introduced plant, completely blan- kets utility poles, abandoned vehicles, and anything else in the way. The Old World climbing fern (*Lygodium microphyllum*) is beginning to do the same in Florida (plate 104). In the garden, species like bindweed, dodder (*Cuscuta*), and many weedy cucurbits follow suit, growing on and over gar- den plants.

Wide environmental tolerances. Species that thrive only in very limited ecological conditions do not tend to become widespread. Some plants, however, tolerate a wide variety of environmental conditions, a trait that enables them to grow over extensive geographic ranges and to occupy a wide diversity of sites; plants in this group may become troublesome weeds.

Weeds also tend to tolerate poor conditions that might not support preferred plants. Purslane (*Portulaca oleracea*), spurges (*Euphorbia*), tumbleweed (*Salsola kali*), and mallows (*Malva*), among others, are drought-tolerant and can outcompete many garden and lawn species under conditions of water stress (plate 105). Purslane can even survive being uprooted and dried out; seemingly dead plants sometimes revive and reroot themselves after a rain or an unintentional watering. Other weeds, including horsetails (*Equisetum*) and comfrey (*Symphytum officinale*), can tolerate anaerobic soil conditions caused by flooding, waterlogging, or severe compaction. Plantain (*Plantago*) tolerates soil compaction particularly well.

Self-compatibility. In many weeds, pollen can successfully fertilize ovules in the same flower or in another flower on the same plant, making them self-compatible. This trait enables seed set under conditions unfavorable for pollination by insects or wind. (Some self-compatible flowers do require an insect or the wind to move pollen, but a receptive flower may be only inches away.) More genetically diverse offspring result from crossing with other plants, and many self-compatible species have mechanisms that minimize the probability of self-fertilization. Weeds that do self-fertilize may compensate for this potential genetic drawback by producing large numbers of seeds—another common feature of weeds. Many members of the pea family (Fabaceae or Leguminosae), snapdragon family (Scrophulariaceae), and mint family (Lamiaceae or Labiatae), including weedy species, are self-compatible.

One extreme example of self-fertilization is seen in flowers that never open, a condition called cleistogamy. Quite a few plants, including most North American violets (*Viola*), produce two types of flowers: the familiar colorful flowers for which they are planted and small, inconspicuous, petal-less flowers that never open. These cleistogamous flowers produce only a small amount of pollen, which is deposited directly onto the stigma within the same flower. Relatively few seeds mature in these flowers, but they augment any seeds produced in the spring. In years with little spring seed production, fertilization of cleistogamous flowers serves a bet-hedging strategy against the possibility of poor seed set. Cleistogamous flowers are particularly common in spring-flowering woodland species. Downy brome (*Bromus tectorum*); ground ivy (*Glechoma hederacea*), a troublesome weed of eastern lawns; and *Commelina benghalensis*, a West African weed, also produce cleistogamous flowers.

Pollination by unspecialized pollinators. Few insect-pollinated weeds rely on only one or a few species of insects. These plants are more likely to be generalists, attracting a large variety of visitors who are also generalists. Although these pollinators may not be as efficient in their transfer of pollen as evolutionary specialists, which learn to work particular flower types, generalists are likely to be present in large numbers. Relying for pollination on many species also provides insurance in the event that one or a few pollinating species are uncommon in a given year.

Weeds often attract pollinators by producing flowers laden with nectar. Umbelliferous species such as Queen Anne's lace (*Daucus carota* ssp. *carota*) and composites such as goldenrods (*Solidago*) produce large numbers of small flowers that on sunny days can be teeming with insects. Milkweeds, wild brassicas (*Brassica*), and species of brambles (*Rubus*) are also heavy nectar producers. Although gardeners may not welcome these particular plants, beekeepers value them highly. Ironweed (*Vernonia fasciculata*) also forms many nectar-producing flowers, but its bitter nectar results in honey with an unpleasant taste.

Wind pollination. Many weeds avoid the need for pollinators by relying instead on the wind for pollen dispersal. Wind-pollinated weeds include most grasses, pigweed, most *Chenopodium* species, and ragweed—not coincidentally, sources of hayfever torment and some of our most noxious weeds. Although wind can transport pollen, it is not an efficient vector, as evidenced by the large amounts of pollen that collect in roadside puddles and on the surfaces of furniture. As a result, natural selection has led to the production of massive amounts of pollen in these plants, far more than in an average insect-pollinated plant. This is bad news for those who have pollen allergies, and another good reason for keeping these weeds out of gardens and yards.

Many seeds. Weeds tend to produce exceptionally large numbers of seeds. A single plant of common mullein (*Verbascum thapsus*) can produce more than 200,000 seeds; and cattails (*Typha*), lamb's-quarters, and purple loosestrife (*Lythrum salicaria*) are alleged to produce as many as a million seeds or more on a single especially vigorous plant. This reproductive strategy maximizes the potential number of offspring, but as a result, the investment of resources in each seed is necessarily limited (that is, seeds are small). Coconut palms (*Cocos nucifera*) and mangoes (*Mangifera indica*) use a different strategy, producing relatively few seeds, but with a much greater investment of resources in each. These extreme approaches to seed production are sometimes contrasted as the "few large or many small" strategies. Although there are many exceptions, weeds tend to fall toward the many-small end of the spectrum.

Long-lived seeds. The seeds of many weeds can survive for years in the soil, while waiting for favorable conditions. Although the life span of a seed depends on genetic as well as environmental factors (for example, soil moisture and temperature), seeds of the evening primrose (*Oenothera biennis*) and curly dock (*Rumex crispus*) have been known to survive in excess of eighty years. In fact, there has been interest lately in germinating seeds of ever-greater antiquity, including successful germination of a sacred lotus seed estimated to be in excess of 1200 years old!

Seeds that survive for more than a year form what is known as a seed bank—a buildup of seeds in the soil in excess of what was produced in the preceding year. Gardeners are sometimes surprised by the reappearance of a species that has been absent for years. Although the seed may have been carried in by the wind or a bird, it may also have been present in the soil, waiting for favorable germination conditions.

Some weedy species produce two kinds of seeds with different longevity and germination requirements. As a result, some seeds produced by a plant might germinate the following growing season, whereas other seeds produced by the same plant might not be capable of germinating for two or more years. Cocklebur (*Xanthium*), for instance, uses this strategy.

Efficient seed dispersal. Although some weeds simply drop seeds at their bases, most disperse seeds widely and efficiently. Dandelions

(*Taraxacum*), maples, milkweeds, and thistles (*Cirsium*) take advantage of the wind (plate 106). Beggar's ticks (*Bidens*), cockleburs, burdock (*Arctium lappa*), and others produce seeds with barbs, bristles, hooks, and hairs that stick to passing animals. Other seeds are consumed by fruit-eating animals; these seeds are later returned to the environment when the animal defecates, often at a considerable distance from the mother plant. Weeds in this group include poison ivy, mulberries, pokeweed, and Japanese honeysuckle. Other troublesome weeds, including jimsonweed (*Datura stramonium*) and horsenettle (*Solanum carolinense*), can be dispersed by moving water. In one study, seeds from seventy-seven weedy species were found in the Colorado River, which is used to irrigate farms and lawns in parts of Arizona and California. There are few home gardens in which water dispersal is a problem; unfortunately, weed species that are dispersed by water are also generally dispersed by other means as well.

One more important vector of weed seeds needs to be mentioned, and that is the gardener. We inadvertently transport unwanted seeds on our garden tools, shoes, and clothing. We may also plant a few unintended seeds that are present as contaminants in the seeds we procure, particularly seeds purchased in bulk or acquired by exchange with private individuals. However weeds reach our gardens, their ability to travel allows them to reach disturbed areas quickly and to get a head start on the plants we would prefer to grow.

Vegetative spread. Some weeds spread vegetatively, either instead of or in addition to spreading by seed. Canada thistle can spread 15 feet (5 m) or more per year through growth of its underground rhizome. And it's not unusual for cinquefoils (*Potentilla*) or wild strawberries (*Fragaria*) to find their way into a garden by sending runners over the ground from where they originally established.

Some weeds store considerable food reserves underground, allowing them to spread effectively or to regrow if their aboveground parts are eaten, damaged, or repeatedly cut back by a gardener. This is true of many difficult-to-eradicate weeds such as field bindweed, common chicory (*Cichorium intybus*), and quackgrass. A gardener who attempts to destroy these plants by using a disc, rotary tilling, or hoeing often succeeds only in dispersing viable pieces of their underground structures, thereby spreading the very plant targeted for eradication. A very different vegetative strategy, one that causes problems in greenhouses, is seen in *Kalanchoe*. This plant produces prodigious numbers of plantlets along its leaf margins. These plantlets then drop into and quickly establish in neighboring pots (plate 12).

Clearly, there are many ways for weeds to compete with other plants. No single weed species has all of the characteristics described above—although gardeners might argue that quackgrass or field bindweed comes close—but a plant that effectively combines several of these traits may pose nearly insurmountable challenges in the garden.

David Quammen described weeds well in an essay published in the October 1998 issue of *Harper's*: "They reproduce quickly, disperse widely when given a chance, tolerate a fairly broad range of habitat conditions, take hold in strange places, succeed especially in disturbed ecosystems, and resist eradication once they're established. They are scrappers, generalists, opportunists. They tend to thrive in human-dominated terrain because in crucial ways they resemble *Homo sapiens*: aggressive, versatile, prolific, and ready to travel."

Where Do Weeds Come From?

A word often heard in association with weed is *introduced*. Introduced weeds are those that originated from outside the area where they now grow. Hundreds of plants that were introduced to North America from other parts of the world are now weedy here. From Asia came kudzu, dock (*Rumex*), jimsonweed, and velvet-leaf. Europe contributed gill-over-the-ground (ground ivy), annual bluegrass (*Poa annua*), and crabgrass (*Digitaria*). From tropical America came species of pigweed and morning glory, and Africa gave us Bermuda grass. And that's to name just a few! However, a plant needn't be introduced to be a garden problem. In fact, some native species are among the worst of North American weeds. A brief sampler for the United States includes horsenettle, pepperweed (*Lepidium virginicum*), ragweed, and poison ivy.

Weeds are not necessarily wild species that have somehow sneaked past our defenses. Many species that are purposely cultivated can become weedy if not properly tended. Crown vetch (*Coronilla varia*) is widely planted along highways and in yards, especially on slopes, but turn your back on it for a season or two and it can spread dramatically, both by seed and by underground stems. Jerusalem artichoke (*Helianthus tuberosus*) is often grown for its nutty-tasting roots or dramatic visual display, but it too can spread by seed and by sprouting roots, especially if adjacent ground is tilled. And many gardeners have had to contend with morning glory, mints (*Mentha*), and trumpet creeper (*Campsis radicans*) as they spread into, around, over, and through garden plantings. Tomatillo (*Physalis ixocarpa*) has become steadily more popular along with Mexican cuisine, but this too can reseed vigorously and can become a troublesome weed in subsequent seasons.

Still other weeds have come not from distant locales but through novel combinations of genes. An example is English cordgrass (*Spartina anglica*), an aggressive weed that has displaced native cordgrass (*S. maritima*) in salt marshes throughout Great Britain, and that is now beginning to spread in North America. *Spartina anglica* does not grow naturally anywhere in the world. It came into existence as a result of hybridization between *S. maritima* (native to Europe) and *S. alterniflora* (native to North America and introduced to Britain). After these two species crossed, the resulting hybrid, *S. ×townsendii*, doubled its chromosome number to produce a new fertile species—*S. anglica*.

Weeds come in many forms, from trees, shrubs, and vines to grasses and forbs. They include annuals, biennials, and short- and long-lived perennials. They are also members of many different plant families, although some families do seem to contribute more than their share to the garden flora we try so hard to eliminate. The worst offending group is probably the grass family, which, in addition to lawn, ornamental, and food crops, includes many, many weedy species. Particularly aggressive grassy weeds include annual bluegrass, crab grass, Johnson grass, and quackgrass.

Some plant families with large numbers of weedy species also include species we go out of our way to plant. The Asteraceae includes not only the troublesome Canada thistle, burdock, dandelion, and ragweed, but also sunflowers, ox-eye daisy (*Leucanthemum vulgare*), asters (*Aster*), and lettuce. The morning glory family (Convolvulaceae) includes not only bindweeds and parasitic dodder, but also domesticated varieties of morning glories and sweet potatoes. The mustard family (Brassicaceae or Cruciferae) challenges us with weedy species of *Brassica* and hoary cress (*Cardaria*), but it also gives us salad radishes and the cole crops (for example, broccoli, cabbage, cauliflower, all derived from *Brassica oleracea*). Other plant families that contribute many of our weeds include the pea family, the buckwheat family (Polygonaceae), the rose family (Rosaceae), and the nightshade family (Solanaceae).

Intraspecific Competition

In the natural world, plants usually do not grow in large, pure stands (monocultures), although there are exceptions. In agricultural fields and gardens or parts of gardens, however, growers commonly plant monocultures (plate 107). Thus, gardeners have an appreciation for the potential effects of intraspecific competition, as evidenced by decisions concerning the spacing of seeds and the thinning of seedlings before plants grow too large. In fact, much of what we know about intraspecific competition has come from observations of what happens when dense plantings of the same species are *not* thinned.

In dense populations, emerging seedlings may perform well for a short time, perhaps as long as their root systems are small and shallow and their shoots do not significantly shade each other. At some point, however, as plants get larger and require increasing amounts of space, nutrients, and water, significant competition begins. Plants can respond to competition in a number of ways, for example, by growing at a slower rate, losing leaves or branches, reproducing at a different age, and changing growth form. But, if density is great enough, plants will suffer eventually and some may even die.

These patterns can be dramatic, especially in species that grow large. Jack pine grows in nearly pure stands on sandy soil across parts of northern North America, including a region measuring 29,000 square miles (75,000 km²) in northern Alberta and Saskatchewan. Young jack pine forests can be so dense that they are referred to as "dog hair," because pushing through these thickets is comparable to what a flea must face. But over time, mortality begins to take its toll, and the forest begins to thin, until eventually what is left is a woodland of scattered mature trees—self-thinning in action. Sooner or later, whether in a jack pine forest or in a too-dense planting of a garden annual, the original large number of individuals can no longer be supported by available resources, and thinning occurs. In nature, the individuals that are eliminated during the self-thinning process are not a random selection of those in the population. Plants that are weak, diseased, unable to reach the light, or that are disadvantaged in any way are more likely to succumb, while those that are larger, healthier, and better defended against pathogens are more likely to survive. This is natural selection in action, and all other things being equal, natural selection should lead to an increase in average fitness in the population.

Gardeners attempt to strike a balance between planting seeds too far apart (in which case a garden may end up with an empty look or low overall yield) or too close together (in which case plants will grow poorly or die). Usually, seeds are purposely placed close together to compensate for failures in germination, among other reasons, and plants are systematically thinned once established. This helps ensure a desired density of healthy, vigorous plants. The ideal density for a specific plant depends on many factors, including its growth form; the texture, fertility, and depth of the soil; weather conditions; and the presence of pathogens and weeds. Plants that can be expected to reach large size should be thinned more aggressively. For this reason it is important to read descriptions on plant labels for the particular cultivars that are being planted. Soil characteristics will also affect the density of plants that can be supported, with deeper, more fertile, and more thoroughly watered soils supporting denser plantings.

Spacing plants with pathogens and weeds in mind might lead a gardener to contradictory conclusions. Providing adequate space between plants can minimize spread of some pathogens, but may encourage weeds, which may need to be controlled in other ways. High-density plantings also tend to magnify drought injury as water becomes limiting. In contrast, crowded plants tend to shelter one another from frost damage.

In general, the better conditions are for a particular species, the greater the density at which growth rate or yield can be maximized. However, it has been shown repeatedly that above a certain density, total yield no longer increases even in the presence of excess resources. If total yield reaches a plateau, then increasing plant density beyond this point means that average plant size must decrease. This fact is particularly important to farmers, who must carefully control their considerable expenditures on seeds, but it is also relevant to backyard growers. If total yield of peppers, for instance, is not increased at high plant density, then a dense planting will yield either very few large fruits or, more likely, a large number of tiny fruits. Even worse, the part of the plant that is harvested may suffer more than the rest of the plant. Above a certain density, parsnips will continue to produce smaller and smaller roots and larger and larger shoots even though total plant yield levels off.

Under the right conditions, however, plants can thrive at surprisingly high densities. In greenhouses, 30-foot (9-m) tomato vines can be grown only 4 inches (10 cm) apart in raised compost beds under constant irrigation and fertilization or by using hydroponic techniques. The aerial parts of the plants are prevented from competing for light by being pruned to single vines, which are tied to cables suspended from the ceiling. As the vines grow, the cables are slowly lowered to the ground, where the nonproductive base of the vine is coiled up on the greenhouse floor.

Intraspecific competition also occurs in other organisms important to gardeners. In years of high population density and food scarcity, deer may be more willing to spend time near houses and in gardens and to eat plants that they otherwise avoid. Insect populations often seem to increase as if competition does not apply to them. However, close observation has shown otherwise. For example, experiments with planthoppers (order Homoptera) showed that as density increased, survival and body length decreased and the time needed for development increased.

Interspecific Competition

Competition is also important among individuals of different species and has a variety of effects. Competition may reduce the population size of each competing species, especially when limiting resources are in particularly short supply. This was seen in a long-term study of two species of Darwin's finches (*Geospiza*) in the Galapagos Islands. Prior to a drought, population sizes were relatively stable, but as a long and severe drought progressed, both populations declined, one severely. In other cases of competition, one species may displace another, as seems to be happening as mallards (*Anas platyrhynchos*) move into wetlands occupied by black ducks (*Anas rubripes*). The invasive plant purple loosestrife (*Lythrum salicaria*) has similarly displaced wetland plant species as it establishes and forms extensive monocultures.

There can be no doubt that gardeners appreciate the potential importance of interspecific competition, given the long hours we spend ridding gardens of weeds. We clearly believe that the presence of these intruders will interfere with the growth and well being of our plantings. Ask a botanist what a weed is, and you find that there really is no agreed-upon definition. Ask a gardener the same question, and you get a quick, simple description of a plant growing where it is not wanted. Ralph Waldo

Emerson described a weed in more poetic terms as "a plant whose virtues have not yet been discovered." (This quote has often been mistakenly attributed to Walt Whitman, who had much to say about plants, especially leaves of grass.)

What one person calls a weed may be purposely planted by another. Cotton growers may rip out morning glories at every opportunity, whereas a homeowner across town may grow these same plants for the beauty and variety of their flowers. Cultural differences may also lead to different conclusions about weeds; burdock, which is considered a nuisance in North America, is a valued vegetable crop in Japan. Many people plant or encourage the introduced St. John's-wort (*Hypericum perforatum*) due to its ascribed medicinal properties, despite its aggressive displacement of native plants. Gardeners come in as many varieties as do the plants in their gardens, and while some pull out anything they did not plant, others are more tolerant, even taking pleasure in watching certain plants that arrive uninvited, perhaps just to find out what they are.

Because there are limited resources in any natural environment, the introduction of one species is often closely followed by the decline or local extinction of one already present. In eastern North America, one of the most notorious introduced plants is purple loosestrife, described by the Nature Conservancy as "beautiful but deadly" and a "purple plague" (plate 108). This admittedly attractive plant, which grows along pond edges and in wet soil, is so effective at eliminating native species that several states have prohibited its distribution. Even so, it is all too often planted in gardens and was only recently removed from the college campus where Steve Carroll teaches.

Two adjacent plants in a garden, even if different species, may have similar water, light, and nutrient requirements. Based solely on an understanding of how competition works, it might be assumed that one or the other of these plants will be eliminated. A garden, however, is not a completely natural environment. Even though two related species or horticultural varieties might not coexist side by side in a natural setting, growers often enable this to happen in the garden. The realities of competition do suggest that if related varieties or species are grown in dense groupings, as in a garden bed that includes many varieties of daffodils, lilies, or hostas, special care may be required, such as providing supplemental water and nutrients. In addition, if two related species share an herbivore or a parasite, growing the plants near each other may make life easier for this pest. For instance, a major herbivore of potatoes, the Colorado potato beetle, was only a minor pest of wild *Solanum* species in western North America until the introduction of cultivated potatoes into their natural range. The potato beetle then added the cultivated potato to its diet and was able to spread worldwide.

Although unwanted species continually find their way *into* gardens, we less often find escapees from our gardens growing in nearby fields or woods. Garden plants produce seeds, many of which are dispersed, but these do not often establish successfully (squash plants on the compost pile notwithstanding). Why is this true? Charles Darwin gave some thought to this question, pointing out the importance of interspecific competition when he referred to "the prodigious number of plants which in our gardens can perfectly well endure our climate, but which never become naturalized, for they cannot compete with our native plants." Even so, as natural ecosystems have become more and more disrupted and as exotics cross international borders at an ever-expanding pace, introduced species have been able to establish in increasing numbers, to the detriment of native species. What might Darwin have to say today were he to reassess the establishment of exotic species in the landscape?

Herbivory

One of the major challenges facing a plant is herbivory, in which all or part of its tissues or fluids are consumed by another organism (an herbivore). In some cases, for example when an insect consumes part of a garden plant without killing it, herbivory is similar to parasitism. In other cases, herbivory more closely resembles predation, as when a kangaroo rat (*Dipodomys*) consumes a seed, or a cutworm devours a seedling, killing that plant as surely as when a raccoon eats a crustacean. Herbivory in the garden takes many forms, including chewing of leaves by caterpillars, tapping of vital fluids by aphids, and consumption of seeds by rodents and birds (plate 109). Whether a garden consists of a few container-grown herbs on the back patio or extensive plantings over many acres, a gardener is certain to encounter herbivory sooner or later.

As often as not, it is the herbivory, rather than the herbivore itself, that is first noticed. Steve Carroll quickly spotted semicircular notches along the leaf edges of a newly planted lilac (*Syringa vulgaris*), but considerable searching was necessary before the responsible party, a caterpillar, was found hiding on the underside of one of the leaves. Likewise, a gardener may deduce from the evidence of decapitated or girdled plants that a rabbit or deer has recently dined in the garden, yet the culprit may never be seen.

In natural terrestrial plant communities, it is estimated that herbivores consume an average of 10–20 percent of the total plant tissue produced. In some environments, though, loss to herbivores can be much higher. In the Serengeti of East Africa, as much as 60–70 percent of plant tissue produced in a year can be consumed by grazing mammals; in Scandinavia, during years of large lemming populations (rodents such as *Lemmus* and *Dicrostonyx*), herbivory can reach 90 percent. Even in North America during outbreaks of pests such as the gypsy moth, spruce budworm (*Choristoneura fumiferana* and *C. occidentalis*), or pine sawfly (*Neodiprion*), individual trees can be completely defoliated. However, herbivores in most temperate regions tend to remove less biomass than this. In a garden or field, herbivory can be especially high if plants are arranged in large monocultures. Under these conditions, herbivores can feed efficiently, because time can be spent eating instead of searching for the next meal.

Another truth about herbivores is that they don't necessarily take turns. At any given time, on any given plant, there are likely to be many kinds of herbivores feeding simultaneously (plate 110). This onslaught may compound the problems of herbivory, as the activities of one herbivore may weaken a plant and make it more susceptible to the feeding of the next—although in some cases, prior herbivory may activate defense mechanisms that enable a plant to more successfully repel later attacks.

Timing and Patterns of Herbivory
The timing of herbivory is critical. Consumption of a seedling will kill a plant instantly. It is all the more unfortunate then, that seedlings, which can be easily bitten in half or swallowed whole, tend to be less well defended against herbivores than are older plants, which often have a combination of chemical and physical defenses. Snails and slugs, in particular, preferentially feed on seedlings. In one study of herbaceous plants, gray garden slugs (*Deroceras reticulatum*) found seedlings more palatable than older plants in 72 percent of the species studied.

Plants that lose substantial foliage early in the growing season, especially well-established perennials, sometimes produce a second flush of growth, which may allow them to produce at least some flowers or a reduced seed crop. But producing a second leaf crop may drain nutrient reserves in the roots, making plants more susceptible to parasites or less able to survive the next winter or dry season. If a perennial suffers foliage destruction late in the season, however, there may be less effect on long-term growth, even though the plant may be unable to produce a new set of leaves that year. However, late-season damage to buds destined to produce next year's growth is likely to cause severe injury.

Patterns of herbivory over long periods of time can also have important consequences for plants. Two plants that lose approximately the same amount of tissue over a several-year period may fare very differently. A plant can probably survive moderate losses to herbivores each year, but a major loss during one year might make a plant an undesirable garden specimen as well as compromise its ability to survive to the next growing season. Although some plants are able to recover fairly quickly from moderate herbivory, this is not always the case. Conifers such as pines (*Pinus*) and arborvitae (*Thuja occidentalis*) tend not to replace tissue during the same growing season in which it is lost. This fact helps explain why tracts of evergreen forest denuded by spruce budworms are so slow to recover. This susceptibility should also encourage a close watch over conifer plantings in the yard and garden.

To an herbivore that feeds on leaf tissue, not all leaves are created equal. Buds and young expanding leaves are more susceptible to herbivory than older leaves and petioles, which are usually tougher, better defended chemically, or both. As a result, insects such as sawflies (family Tenthredinidae) and gypsy moth larvae often feed on young leaves and ignore older leaves (plate 111). Mammals may also feed on younger, more nutritious, and more palatable plant parts—as was the case when white-tailed deer (*Odocoileus virginianus*) selectively ate the tender tip-leaves and young okra pods in Steve Salt's gardens. It is wise to inspect and protect garden plants early in the growing season, when they are especially tender and inviting, and to inspect young plant parts periodically through the growing season. If you are seeing significant tissue loss without finding an obvious perpetrator, plants should be inspected at dusk or after dark. Rabbits and deer are more likely to be found in the garden at that time, as are nocturnal feeders such as cutworms and slugs.

Types of Herbivory and Their Effects

Herbivores come in many shapes and sizes, and they feed on plants in a startling variety of ways. They may feed on aboveground vegetative tissue, such as leaves and stems; reproductive tissues such as flowers, fruits, or seeds; or less conspicuously on roots. The tissue herbivores feed on, the amount of tissue they remove, when during the growing season they feed, whether they transmit pathogens, and many other factors ultimately determine how drastic an effect an herbivore will have on a particular plant.

Grazing and browsing. We can loosely define grazing and browsing as consumption of the aboveground tissue of herbaceous and woody plants, respectively. Grazing and browsing can affect plant growth and reproduction to different degrees. If a woodchuck (*Marmota monax*) eats the top off of a sunflower just before its single flower head opens, the result is rather drastic. However, if insects reduce a plant's leaf area by a few percent, the impact may be minimal. Grazers and browsers include some of the most familiar garden herbivores, ranging from butterfly larvae to slugs to deer.

Removal of stem and leaf tissue by herbivores (or by gardeners with pruning shears) can dramatically affect plant architecture (growth form). Herbivores that eat or destroy a plant's terminal bud can have an especially profound effect on plant shape. Removal of terminal buds creates lower, more spreading plants with a greater number of fruiting or flowering side branches. This may be a desired effect, as when growers of peaches, cherries, and chrysanthemums pinch out terminal buds. Removal of apical buds can also be used to grow more densely leafed hedge plants and to help correct legginess of indoor plants resulting from inadequate light. But excessive bushiness of tomato plants, for instance, can lead to delayed fruit ripening and increased fungal disease. The loss of apical dominance may be especially disastrous for growers of conifers. One summer, nesting blackbirds at a northern Missouri tree farm damaged the terminal buds of hundreds of conifers. This caused these potential Christmas trees to produce a rounded top rather than the hoped-for pyramid-shaped apex.

The growth patterns described above are true for many plants, but they are not universal. Grasses grow from thickened nodes along the length of the stem, as well as from the base of stems. Cell division at these nodes causes growth from below, rather than from above, and it allows members of the grass family to tolerate relatively severe grazing as long as basal parts are not injured (plate 112). As a result, moderate grazing from above does not necessarily destroy a plant's ability to continue growing normally. Understanding this growth pattern is relevant to gardeners who grow the increasingly popular ornamental grasses or wet-site plants such as horsetails, which also have meristems along the stem.

There are other consequences for herbivory of growth from below. Because the upper portions of grass blades are older than lower portions, the upper parts of stems are usually tougher and may be less efficient photosynthetically than lower parts. Because it is a form of thinning, grazing of grasses may allow more light to reach lower, more photosynthetically efficient parts of the plant. But if grazing is too intense, is repeated too frequently, happens late enough in the growing season, or occurs too close to the base of blades (as is often the case when slugs feed), a grass plant may be prevented from flowering and producing seeds.

Trees are susceptible to an assortment of leaf-feeding herbivores. Especially notorious species include the gypsy moth and the Japanese beetle, which have caused immense damage to garden and yard plantings and native forest trees. As unlikely as it might seem, given their large size, protective bark, and heavily reinforced woody tissue, it is sometimes surprisingly easy for herbivores to damage or kill a tree. This is true because in most trees, the vascular cambium (which enables radial growth) and the phloem (which transports carbohydrates) are just beneath the bark (plate 113). Thus, a tree can be killed if the vascular cambium and phloem are severed completely or nearly so. This is what happens when a tree is girdled, whether by a beaver (*Castor*), a forester, or a homeowner repeatedly ramming a young tree with a lawnmower. Mice and deer can also severely injure trees by gnawing on bark. In northern climates, girdling damage of fruit trees is sometimes discovered after snowmelt, evidence that mice have been feeding from the protection of their tunnels. North American porcupines (*Erethizon*) also girdle trees, although they usually do not completely encircle trunks as they feed.

Sucking insects. In contrast to damage caused by grazing, which may be visually dramatic, the feeding of sucking insects may go unnoticed unless plants are examined closely. Aphids are the best known of the sucking insects. They feed by inserting their tiny stylets into a plant's phloem tissue and extracting the sugar-containing solution produced by photosynthesis.

(This way of feeding can be compared to the way mosquitoes feed on us.) Aphids generally congregate at the growing tips of shoots, where plant cells are less heavily armored. This strategy makes insertion of their stylets easier and gives them access to actively growing tissues that are receiving fluids relatively rich in sugar and nitrogen.

Although most aphid species feed on only one or a few closely related plant species, some feed more widely. There are also aphids that switch host species partway through the growing season. Given their small size, however, aphids can sometimes go unnoticed, especially those that are green; black or red aphids or aphids that are being tended by ants are more easily noticed. If aphids are seen on one plant in the garden, it is a good idea to check other nearby plants as well. However, there are many insect species that feed on aphids, including lacewings and ladybird beetles. Wise gardeners encourage populations of these predatory species—the so-called beneficials—to make other control measures less necessary.

Thrips (order Thysanoptera) also tap into plant juices, and they rasp and scar flowers, fruits, and leaves. Spittlebugs (family Cercopidae), which are evident from the protective froth that surrounds their young, tap into xylem cells. Spittlebugs can be quite numerous; in one study conducted in a Michigan grassland, the common meadow spittlebug (*Philaenus spumarius*) was the most abundant insect (by weight). Whiteflies are particularly common indoors and in greenhouses. These insects can often be seen resting on leaves and rise as a cloud when disturbed. Squash bugs, too, feed on plant fluids and can be devastating pests of pumpkins, squashes, and other cucurbits. In fact, a severely infested patch can be so sucked dry of phloem as to appear to have been burned (plate 114).

It can sometimes be difficult to assess the impact of aphids and other sucking insects. Small populations may not inflict visually obvious damage, although overall plant vitality and productivity may decline. But severe infestations, especially if apical meristems are damaged, may produce mottled, twisted, or shriveled young plants that remain deformed for life or may even die. If aphid numbers become too great, the honeydew they produce (a sugar-rich exudate) can accumulate on lower leaves and cause discoloration as a charcoal mold begins to grow. Aphids can also harm woody plants. In a study of lime trees (*Citrus aurantifolia*), saplings with and without aphids did not differ in plant height or diameter, leaf number, or leaf size. But aphids did have a significant effect on root growth, and ultimately, on total sapling weight. Yet another significant negative effect of sucking insects is their role as vectors of numerous viral and bacterial diseases, as described in the section on parasitism in this chapter.

Mining and boring. Some herbivores feed from within plant tissue by mining or boring. Leaf miners are common and are easily recognized by the characteristic tunnels carved within leaf tissue. They damage vegetables such as peas and spinach as well as ornamentals and many native species, and they pose an especially serious threat to seedlings. Leaves with evidence of tunneling should be removed and destroyed, because the responsible party may still be inside.

Gardeners who grow pumpkins and zucchini may be familiar with squash vine borers (*Melittia satyriniformis*). These moth caterpillars bore into the lower portions of stems, where they can sometimes be discovered by what appears to be sawdust (actually frass) accumulated at the base of the plant. Borers can cause sudden wilting of squash plants. They are sometimes controlled by slitting the stem and extracting the grub or by injecting the insecticide *Bacillus thuringiensis* (Bt) with a hypodermic needle, as described in chapter 6. Numerous species of beetles excavate tunnels in the wood of trees and shrubs, often causing more damage by the wood-rotting fungi that they spread than from the tunneling itself (plate 115).

Root herbivory. The most familiar herbivores are those that feed aboveground, because it is there that we are likely to notice them. But there are also many herbivores that feed on roots. We are most likely to encounter this group when we cultivate the soil or when we notice evidence of their tunneling, although sufficiently severe feeding on the roots will eventually result in symptoms in aboveground tissue, such as wilting, even with adequate soil water.

Cicadas are best known and most often noticed in their noisy adult stage. Although they damage young twigs during egg-laying, cicadas' major damage comes not as adults but during their long larval stage, during which they feed on plant roots. Root-feeding Japanese beetle larvae may do as much damage as the leaf-chewing adults, and beetle larvae (grubs), in general, are root herbivores. When a lawn is converted to a garden, many grass-feeding grubs may be present in the soil, and it is a good idea to turn the soil over thoroughly after removing sod. Any grubs that are seen can then be removed. It is also helpful to let the ground lie fallow for at least one year or to grow annuals that are not susceptible.

Nematodes may be the most severe root pests worldwide. They damage plants by puncturing the walls of root cells and sucking out cellular fluids, which can lead to lesions, galls, or excessive root branching. Infested plants may wilt, grow poorly, and even die. Commonly affected garden plants include beets, tomatoes, sweet potatoes, and many ornamental flowering species (plate 116). Nematodes are especially devastating in certain tropical and subtropical sandy soils, where they reportedly cause more damage to plants than do insects and plant diseases combined.

Common burrowing mammals can damage roots of garden and lawn plants. Pocket gophers (family Geomyidae) feed on roots, bulbs, and tubers as well as on grasses, seeds, and the bark of trees such as cherries. Ground squirrels (*Spermophilus*), moles (family Talpidae), and other burrowers do not usually eat roots in significant amounts, but can damage plants by extensive burrowing.

Forming galls. One of the most interesting ways in which invertebrates feed on and use plants is by causing plants to form tumorlike galls. Produced in response to invading organisms as diverse as bacteria, mites, wasps, and flies, galls form when the host plant is stimulated to produce a mass of unorganized cells (plate 117). This can occur when an invader produces plant hormones or hormone mimics or when genes that regulate hormone production are transferred from the intruder to the plant. Most galls form on broad-leaf plants. They vary from inflated growths on the stems of goldenrods, through wartlike growths that can virtually cover the upper surface of cherry leaves, to reddish swellings that form on grape vines (caused by flies, mites, and midges, respectively). These growths provide shelter and food for the inhabitant; if no exit hole is present, the perpetrator can often be found inside.

Although not all galls cause serious damage, some injure plants through mechanical weakening of stem tissue, especially in woody perennials. Fruit trees, shrubs, and canes, including plums (*Prunus americana, P. domestica, P. salicina*), blueberries, and raspberries, are particularly susceptible to crown gall, caused by bacteria. Pruning offers the best means of treating infected plants.

Seed and fruit herbivory. There's an old joke that goes, "What's worse than finding a worm after biting into an apple? Finding only half a worm." As we know, then, herbivores also feed on seeds and fruits.

Some herbivores consume seeds that are still on the plant, for example, slugs that feed on ripening tomatoes and the seeds within and birds that feed on sunflower heads. Others consume seeds on or under the soil surface, for instance, the brown-headed cowbird (*Molothrus ater*) that devastated one of Steve Salt's ornamental corn plantings. There are even species, particularly weevils, whose larvae develop inside seeds and feed from the inside out; this is a major problem in stored grains (plate 118). Seed-infesting weevils can present particular problems for gardeners who save seeds of rare or heirloom plant varieties. Fortunately, seeds purchased from reputable seed venders rarely harbor weevils. Home-saved seeds may be protected from weevils by storage in freezers after sealing the seeds in moisture-proof containers. Even if weevil eggs are present, they will not hatch under these conditions, although they may lie dormant and then hatch when seeds are returned to warmer conditions.

It is difficult to assess the extent and importance of seed predation in nature. In gardens, we make decisions about what plants to grow and where to grow them, and we monitor what happens to our seeds once planted. Therefore, significant seed predation in the garden is likely to be noticed— although the act of predation may not be observed. This is especially true if a seed packet is discovered to be full of weevils or grain moths, if raccoons or birds clean out your berry patch, or if blackbirds follow along as you plant corn/maize, eating your seeds only moments after they are planted.

Fruit herbivory is common and frequently dramatic, both in nature and in the garden. But unlike seed predation, which benefits the predator but is lethal to the plant embryo contained within, fruit consumption often benefits both the herbivore and the plant. This is true because at least some of the seeds within the fruit may survive the experience. A flock of Bohemian waxwings (*Bombycilla garrulus*) might consume most of the fruits on a chokecherry (*Prunus virginiana*), but many of the seeds will be viable after they pass through the birds' digestive systems and may be dispersed some distance from the parent plant. Similarly, squirrels gather and eat many acorns, but some of those they bury for later consumption are forgotten and may then germinate to produce an oak tree.

Pollen and nectar feeding. Many organisms, including birds and mammals but especially insects, take advantage of the rich food source offered by flowers full of pollen and nectar. By consuming great amounts of pollen, pollen feeders can sometimes damage flowers or keep a plant from reproducing successfully. For example, pollen-feeding blister beetles (family Meloidae), cucumber beetles, and bean beetles (*Acanthoscelides obtectus*) may consume enough pollen to prevent fruit production. Cucumber beetles pose a double-edged problem because their larvae feed on the roots of corn/maize and other plants—hence, their other common name of corn rootworm.

However, many pollen and nectar feeders provide a service to plants through their pollination. Best known among the "good" insects are bees, especially honey bees and bumble bees (plate 119). Although bees consume some pollen, they also transfer prodigious amounts from one flower to another, usually with little damage. As a result, they are generally beneficial visitors to our gardens and yards—even if we do occasionally get stung. As is true of fruit, nectar attracts herbivores that are rewarded for their efforts, in this case with sugar- or nutrient-rich nectar. These usually then pollinate the plant in return.

Why Is the World Green?

With so many herbivores loose in the world, it might seem a miracle that any plants survive to the end of the growing season. Yet, except in regions such as extreme deserts and salt flats, the overriding impression one has of the landscape is that it is green. How is this possible? Although research and debate about herbivory have taken many turns, there are two central arguments explaining the plant world's apparent success. The first is that herbivores, in spite of their considerable diversity and numbers, also have large numbers of effective and often deadly enemies. As a result, the ultimate detrimental effects of herbivores are minimized. (We discuss the role of predators and parasites at length in the next two sections of this chapter.)

The second argument for why the world is green centers on the plants themselves. Because plants are rooted in place and unable to move, at least in the conventional sense, it's easy to think of them as hapless victims in the face of their enemies. However, plants possess a startling array of chemical and physical defenses against herbivores. And chemical and physical defenses are not mutually exclusive. Many plants have specialized hairs, the tips of which release toxic chemicals when broken, as anyone who has brushed against stinging nettles (*Urtica dioica*) can testify. In nettles, the stinging hairs are barbed, making them even more likely to catch on a trespassing animal. Furthermore, cell walls at the tips of these hairs are impregnated with silica, making them glasslike and more likely to break. As a result of their physical and chemical defenses, plants are a surprisingly poor food source.

Physical defenses. Plants have obvious physical defenses against herbivores in structures such as the spines, prickers, and thorns found in cacti, thistles, nettles, and thousands of other plants. Herbivores must expend more time and energy overcoming defenses of thorny plants than those of less well-armored plants. With exceptions such as roses and blackberries, our common cultivated plants do not have such physical defenses, for they would be as likely to impale us as they would a hungry herbivore. Many garden plants do have defenses, however, such as dense tangles of bristles or sharp stiff hairs, which may seem inconsequential to us humans, but which can present formidable barriers to slugs, insects, and other small herbivores (plate 120). Consider, for example, the hairy stems of sunflowers, okra, and upland cotton (*Gossypium hirsutum*, in which *hirsutum* means "hairy"). In addition to impaling or injuring a feeding herbivore directly, structures such as dense hairs may also prevent small herbivores from moving around freely on the plant.

Plants also have less obvious physical defenses. Silica, the mineral of which glass is made, is common in the epidermal cells of horsetails and many grasses. It may both slow the rate of chewing and wear down herbivores' teeth and abrade digestive tract linings. In addition to regulating transpiration, waxy compounds on leaf surfaces can provide a barrier between feeding herbivores and the leaf's epidermal cells. Especially valuable tissues such as terminal meristems in shoots are protected from small herbivores, as well as from the environment, within resistant bud scales.

Seeds and fruits are also defended physically. Although not usually covered in spines, many seeds possess a hard seed coat that may require great effort or a long, tedious process to chew or crack open. Many fruits are resistant to herbivory by virtue of thick impenetrable fruit walls, waxy coverings, and an array of hairs and spines. Dramatic examples include the coconut palm (*Cocos nucifera*) and the wood apple (*Aegle marmelos*) of

India, whose wall is so tough as to require a saw or chisel to cut. Other examples include hickory nuts (plate 121), acorns, and the outer rinds of *Citrus* fruits.

Chemical defenses. Plants also defend themselves chemically against herbivores, and they usually rely on more than one type of defensive chemical. A combination of chemicals may act on a wider array of potential herbivores or provide a more potent defense than the effects of individual compounds. Defensive chemicals may be spread throughout the plant or they may accumulate in special glands or tissues (plate 122). Some plant chemicals, such as those with noxious odors or flavors, decrease the likelihood that an herbivore will feed, as a nasty taste will certainly deter a second bite. Other plant compounds may inhibit herbivore maturation, slow down herbivore growth rates, or disrupt reproduction in the herbivores that feed on them. Ageratum (*Ageratum houstonianum*), balsam fir (*Abies balsamea*), and yew (*Taxus*) produce chemical mimics of insect developmental hormones that cause premature molting in insects that feed on them. Lupine and alfalfa produce hormone mimics that affect reproduction in mammalian herbivores.

One of the best-studied chemical defenses is that of white clover (*Trifolium repens*), some strains of which produce sugar compounds that are chemically bound to cyanide. When tissue in this clover is damaged, two enzymes are released that cleave the sugars and release deadly free cyanide into an herbivore's digestive system. Seeds of most members of the rose family, including apricots, almonds, and black cherries (*Prunus serotina*), also produce cyanide.

Another plant chemical defense involves production of pheromones. Pheromones are chemical compounds produced by one individual as a means of communicating with others of the same species. For instance, a wild species of potato (*Solanum berthaultii*) produces a component of an aphid alarm pheromone that causes aphids to take flight.

Categories of plant defenses. One scheme for categorizing plant defenses distinguishes between those that are present in tissues at all times (constitutive) and those that are synthesized in response to herbivory (induced). Although this concept is usually applied to chemical defense, it can also apply to physical defense. In raspberries, for instance, prickles on plants that have been grazed by cattle are longer and sharper than those on plants that have not been grazed. *Acacia* and stinging nettles show a similar response.

Constitutive chemical defenses include compounds such as tannins, which are common in bracken ferns (*Pteridium aquilinum*), tea plants, and oaks. Tannin is known to deter feeding by herbivores ranging from insects to large grazing mammals of the African plains. Humans have learned to leach tannins from plants, making the acorns of many oaks edible, but most herbivores have not learned this trick. Squirrels (*Sciurus*), bears (*Ursus*), deer, and wild turkeys, however, eat large numbers of acorns in spite of their chemistry. These and other herbivores are able to include acorns in their diets by selecting acorns of species or even individual trees having lower tannin concentrations, by mixing acorns with other foods, and through physiological mechanisms that decrease the toxicity of the tannins.

Constitutive defensive compounds can make up as much as 60 percent of the dry weight of some leaves. Even so, plants can be induced to produce even greater amounts of these compounds by herbivory, such as when oaks produce a second flush of leaves, even richer in tannins, following defoliation by gypsy moths. In a similar way, paper birch (*Betula papyrifera*) increases resin concentrations in new shoots after browsing by snowshoe hares (*Lepus americanus*), and grasses increase silica concentrations following artificial defoliation.

It's expensive for a plant to invest resources in defenses that may never be used; and resources thus spent cannot be used for growth and reproduction. An alternative strategy for plants is to produce defensive chemicals only when needed. Tomatoes and sugar beets produce a great variety of induced compounds, and cabbage has a particularly interesting response to herbivory. When fed on by caterpillars of the large white butterfly (*Pieris brassicae*), cabbage releases volatile compounds that attract a braconid wasp (*Cotesia glomerata*), which then parasitizes the caterpillar. There is also evidence for induction in one plant caused by herbivory of another plant nearby. This phenomenon, which is known in willows and alders, is likely a response to a volatile chemical released by the damaged plant.

The time necessary for the induction of chemicals can vary dramatically. If induction requires the replacement of tissue, the response may take weeks or months, or it may not occur until the next growing season. Other induced responses, such as the synthesis of chemical compounds, can occur in hours or even minutes. Although induced responses can be effective and may benefit plants in the long run, most do not reduce herbivory immediately because the stimulus is the feeding itself. As a result, perennial garden plants that use this defense may grow poorly in their first year, and annuals may produce fewer flowers or fruits if they must direct resources to the production of physical and chemical defenses.

Another method of categorizing defensive compounds depends on how abundant they are in plant tissue. Compounds that are present in relatively large quantities are quantitative defenses, as in oak tannins, whereas those that are effective even though present in small amounts are called qualitative defenses. An example of a qualitative defensive compound is atropine, present in the deadly nightshade *Atropa belladonna*. Atropine is a highly toxic chemical related to cocaine that was used under the name "belladonna" by nineteenth-century women to dilate their pupils, which they believed made them more attractive. Other qualitative compounds include caffeine, found in beans of the coffee plant and leaves of the tea plant, and strychnine, present in seeds of the strychnine plant (*Strychnos nux-vomica*). Some qualitative compounds are so toxic that plants must store them in special glands or tissues in order not to poison themselves; others are converted to toxic forms only after an herbivore ingests them. Individual plants can use both quantitative and qualitative compounds or can switch from one type of defense to the other. In sorghum (*Sorghum bicolor*) and black cherry, the concentration of cyanide-generating qualitative compounds decreases as the growing season progresses, while concentrations of the quantitative compounds lignin (discussed below) and silica increase.

As mammals, we gardeners must also contend with the chemical defenses in the plants that we eat. One of the ways we have addressed this problem is by breeding plants having lower concentrations of harmful or distasteful chemicals. Unfortunately, such plants are also more palatable to herbivores. This poses a dilemma for plant breeders, who seek to develop varieties that are naturally resistant to herbivorous pests, yet safe for humans, who, from a plant's point of view, are also herbivorous pests.

Physical and chemical defenses do not come cheaply. Although defenses benefit plants, there is a cost to producing, storing, and mobilizing chemical compounds and to building and maintaining physical structures.

Ecologists have come to consider defenses against herbivores in terms of cost-benefit trade-offs, much as a gardener might weigh the costs and benefits of using a chemical pesticide. A plant has a finite amount of energy to invest in its many functions (growth, reproduction, storage, defense), and energy invested in defense will not be available for other needs. This has been seen in a study of the tropical shrub *Psychotria horizontalis*; plants that produced tougher leaves and higher concentrations of tannin (that is, those that invested more heavily in defense) also grew less. Resources invested in defense must also be partitioned into different tissues. As a result, if defensive chemicals are moved from roots to leaves in response to leaf-feeding herbivores, the roots then become more vulnerable.

A plant's ability to allocate resources in response to herbivory is influenced both by its genetic makeup and its general condition. There is considerable evidence that seriously stressed plants—whether from disease, water deficit, or other environmental factors—are more susceptible to herbivory. This points out the importance of wise plant selection and good plant care.

The poor-food defense. The extent to which plants are eaten is affected not only by spines, tannins, and qualitative toxins, but by the basic ingredients of which plants are made. Aside from water, most plant tissues are composed primarily of carbon-rich, nitrogen-poor compounds such as cellulose and lignin. These compounds can make up as much as 90 percent of the dry weight of plants, and they are low in nutritive value and difficult to digest. Breakdown of cellulose requires the enzyme cellulase, which most organisms do not produce. Lignin is even more difficult to break down, and it generally passes through undigested. Tannins form complexes with proteins that further reduce the quality of plants as food. Overall, plants tend to be low in total protein, salt (hence the craving of many herbivores for salt licks), essential amino acids, and, often, essential fatty acids and fat-soluble vitamins.

The low concentration of nitrogen in plants can be especially important to herbivores—and to gardeners. Nitrogen fertilization may make your plants greener, but it also increases herbivory. Scientists have found better survival and reproduction in insects that feed on nitrogen-fertilized plants than those that feed on unfertilized plants. Furthermore, nitrogen is not uniformly present in all plant tissues or at all stages in plant growth.

On average, nitrogen is more concentrated in young leaf tissue than in older leaf tissue, and seeds are by far the richest in nitrogen. To obtain sufficient nitrogen, as well as other nutrients scarce in plants, herbivores must often consume large amounts of tissue and many preferentially feed on the more nitrogen-rich tissues. Now, consider that gardens are islands of highly fertilized and carefully tended plants in a sea of lousy food. Is it any wonder that there always seem to be more insects feeding on garden plants than those growing outside the garden?

Mimicry as defense. Another way plants avoid herbivory is through mimicry. Many butterflies lay their eggs directly on host plant tissue, which is then fed on by their larvae after they hatch. Females of some butterfly species will avoid a plant if they detect that another insect has already laid its eggs there. This behavior, which may have evolved as a means of reducing competition among the young that emerge from the eggs, opens up a remarkable way for plants to fool egg-laying insects. For example, the passion flower *Passiflora auriculata* produces glandular growths resembling zebra butterfly (*Heliconius*) eggs at its leaf bases, and female zebra butterflies are less likely to lay eggs on leaves that have these growths than on leaves that do not. Another example of mimicry is known in an Australian mistletoe (*Dendrophthoe*). Individuals of this parasitic plant produce leaves whose appearance mimics the leaves of whichever of three host plants it is living on. This makes the mistletoe less conspicuous to its herbivores.

Herbivore Countermeasures to Plant Defenses

As mentioned above, herbivory is a problem in the garden because many chemical and mechanical defenses have been purposely bred out of garden plants. Also, herbivores have an arsenal of strategies at their disposal to help them overcome plant defenses. Indeed, relationships between organisms and their antagonists can be viewed as an arms race, with plants continuously evolving better defenses and herbivores evolving countermeasures against these new defenses.

Dealing with plant chemistry. Many herbivores can detoxify or resist plant chemical defenses. For example the bruchid beetle *Caryedes brasiliensis* is able to detoxify the amino acid L-canaline produced by the tropical legume *Dioclea megacarpa*. Other insects, such as certain caterpillars, have biochemical mechanisms and pathways, such as maintaining high gut pH and sequestering compounds in specialized glands or tissues, that reduce the effectiveness of plant defensive compounds. For example, the large milkweed bug (*Oticopeltus fasciatus*), which ingests cardiac glycosides from its host plants, sequesters these compounds in its own tissues. Many insects produce a group of enzymes (mixed-function oxidases, or MFOs) that can neutralize both generalized and specific toxins. Lepidopteran larvae that feed on many different plant species generally have higher MFO activity than do larvae that feed on only one or a few different plants. This capability in insects has turned out to be of great importance because it also contributes to the development of resistance to pesticides.

In fact, some herbivores incorporate plant chemical defenses for their own protection. Perhaps the best-known example of this turn-about strategy involves monarch butterflies and milkweed plants. Monarch larvae feed on the milkweed, which contains cardiac glycosides (toxins that affect the heart muscle). As they feed, the larvae store the glycosides in their bodies as protection from their own predators (plate 123). Monarchs advertise this toxicity through bright colors that make them stand out. One photograph frequently used in biology textbooks shows a blue jay (*Cyanocitta cristata*) vomiting after eating a monarch, convincing evidence that chemicals can have dramatic effects on feeding behavior.

A tactic used by certain beetles and lepidopteran caterpillars is severing plant secretory canals that contain and deliver toxins. The insects then feed on tissue that would otherwise have been protected. Squash beetles (*Epilachna borealis*), which feed extensively on members of the cucumber family, often cut circular trenches that isolate portions of the leaf. This prevents movement of the defensive chemical cucurbitacin to the tissues being fed on. Cabbage loopers (*Trichoplusia ni*) cut similar trenches across cucumber and lettuce leaves, and then consume tissue on the side of the trench away from the stem. Other insects simply feed between secretory canals. Larvae of leafroller moths in the genus *Agonopterix* roll up leaves of host plants such as parsnip, thus shading parts of the leaves and preventing the necessary light-activation of defensive compounds.

Specialist herbivores—those that eat plants of a relatively few species, genera, or families—are much more common than generalists. Although specialization is uncommon in mammals, well-known exceptions include giant pandas (*Ailuropoda melanoleuca*) and koalas (*Phascolarctos cinereus*), whose diets consist almost entirely of bamboo and *Eucalyptus* leaves, respectively. Eucalyptus leaves are low in nutrients and have high concentrations of lignin, tannin, and other compounds that reduce digestibility. The koala is able to survive on such a poor food source because of adaptations of its own. First, it has a long gastrointestinal tract relative to its body size, allowing more complete digestion of food. In particular, it has a long cecum, a pouch at the junction between the small and large intestines, in which food is held for several days while it undergoes bacterial fermentation; this contributes to breakdown of tough eucalyptus leaf fibers and more complete extraction of nutrients. The koala also has a slow metabolic rate, and it feeds and chews slowly and meticulously.

Not all herbivores, however, are as restricted in their choice of food plants as monarchs and koalas. By feeding on several or many different plants, perhaps from different families, an herbivore such as a grasshopper or woodchuck may avoid accumulating dangerous levels of any one toxin. One price of this smorgasbord approach to feeding, however, may be reduced efficiency in obtaining nutrients from the diet.

An interesting strategy for increasing the efficiency of extracting nutrients from their diet is known in rabbits and some rodents, which re-ingest some of their fecal material (a feeding behavior known as coprophagy) to obtain nutrients that were not extracted the first time through. Perhaps because of this, rabbits are one of the most nutritionally efficient herbivores, reportedly requiring as little as 2 pounds (0.9 kg) of plant protein to produce 1 pound of rabbit protein. In contrast, even ruminants such as sheep require approximately 6 pounds and cattle 8 pounds; notoriously inefficient horses require a great deal more.

Getting physical. In addition to countering chemical defenses, herbivores must also deal with physical defenses. The silica-rich tissues of grasses, for instance, wear down the tooth surfaces of vertebrates. Grazers such as the horse, which feeds largely on grasses, have evolved molars (grinding teeth) with high crowns and multiple ridges of enamel. Nevertheless, the adage "Never look a gift horse in the mouth" refers to the fact that the age (and thus value) of a horse can be determined by the extent of wear-and-tear on its teeth. The teeth of a rabbit continue to grow throughout the animal's lifetime, thus replacing tissue lost to abrasion. Birds, which lack teeth, crush plant tissue with muscular, gravel-filled gizzards.

Herbivores must sometimes deal with defenses that do not interfere with feeding directly, but that interfere with the manipulation of, or movement across, plant parts. Some caterpillars spin a silken web over leaf hairs, allowing them to move more easily across leaves. Slugs and snails lay down viscous slime trails that coat and press down sharp hairs on plants such as tomatoes and potatoes.

Timing. Herbivores can deal with toxic compounds by avoiding them in time rather than in space—by feeding during times when plant tissues are less well protected. For example, insects that feed on oaks tend to be more abundant during the spring than later in the growing season, when leaves are tougher and better protected. In a study of winter moths (*Operophtera brumata*), an herbivore of English oaks (*Quercus robur*), larvae that were fed oak leaves produced in the early spring reached an average weight three times greater than larvae fed leaves produced only a few weeks later, when leaf protein content was lower and tannin concentration was higher.

Mutualism and herbivory. Although herbivory most obviously involves herbivores and the plants they eat, there are many plant-herbivore interactions that include additional species involved in mutualistic relationships with one or the other of the main participants. In a mutualistic relationship (discussed in greater detail later in the chapter), all participating individuals gain a net benefit. One of the best-known defenses against herbivores based on a mutualism is that between ants and acacias (*Acacia*). In this relationship, ants vigorously attack insects and even larger organisms that land on, feed on, or even brush against their home plant. In return, the ants use the plant as a source of food and shelter.

Herbivores are also sometimes partners in mutualistic relationships that help them overcome plant defenses. For example, ruminants such as domestic cattle possess a multiple-chambered stomach containing bacteria and protozoans. These microorganisms benefit from the warm, moist, dark environment provided by their hosts; in return, they break down resistant molecules such as cellulose. Some gut inhabitants minimize the effects of toxic plant qualitative defenses. A number of invertebrates, including termites (order Isoptera), also harbor microorganisms that aid in digestion.

Predation

We know a great deal about the growing requirements of many garden plants, the types and impacts of garden herbivores, and even the physical garden environment, but we often know little about garden predators beyond which species are present and some of the prey that they consume. Gardeners have long believed that predators such as ladybird beetles help keep herbivorous insects in check, but these assumptions are based as much on anecdote as on supportive data, although the evidence for efficacy is strong in the case of predatory insects used for pest control in confined areas such as greenhouses (plate 124). The praying mantis, although almost universally revered by gardeners as a consumer of harmful insects, eats beneficial insects as well, and it is not at all clear that the balance always tips in favor of garden plants. There's a great deal more for us to learn about predation in the garden.

At its simplest, predation is an interaction in which one organism consumes another. In this section we address classical predation (that is, carnivorous predation), the capture, killing, and consumption of organisms other than plants and fungi. But even this narrow definition of predation includes some paradoxes. For example, although the definition excludes animals that feed on plants, it includes plants, such as the Venus flytrap, that capture and consume animals.

Adaptations in Predators

Classical predators, which include everything from leopards to American robins to yellow garden spiders, obtain food in a remarkable variety of ways. Even so, predation—including that in the garden—can be thought of as taking place in three steps: detecting, capturing, and handling prey.

Detecting prey. For those of us whose image of predation is a pride of lions hunting on the African savanna, it's easy to assume that locating prey is the least of a predator's worries. For many predators, however, finding prey may be more energy-demanding than the capture itself. A bat flying across the yard at dusk expends a great deal of energy as it sweeps about in search of flying insects (plate 125), as does a predatory nematode moving through the dark, convoluted passageways of the soil.

Of the many ways in which predators detect prey, visual predation is perhaps the easiest to appreciate, given human reliance on eyesight. Visual predators are widespread in the natural world and include species as varied as peregrine falcons (*Falco peregrinus*), dragonflies, and lizards (plate 126). Airborne predators are especially likely to rely on vision. Visual predators detect and interpret stimuli such as movement, color contrast, and irregularities of outline. Some predators can use wavelengths of the electromagnetic spectrum not detectable by humans. Many insects can perceive ultraviolet light, although this may be more important in pollination than in predation. Snakes in the pit viper family (for instance, rattlesnakes) have sensory pits that are used to detect infrared energy. This allows the snake to find warm-blooded animals, especially at night or in dark burrows, by detecting the heat that they radiate. Snakes with two sensory pits are thought to have some depth perception in the infrared.

Auditory cues are also widely used by predators. By turning its head from side to side, an owl efficiently receives sound waves that aid in prey location. One of the most remarkable adaptations of auditory hunting is found in bats (order Chiroptera) and porpoises (family Phocoenidae), which produce high-frequency sound waves that bounce off objects in the environment. This means of hunting, called echolocation, is extremely effective when feeding in air and water. By detecting how much time a pulse takes to return to it, a bat or a porpoise can estimate distance, and by processing differences in these times between right- and left-ear reception, it can establish direction. Some bats can detect and locate prey that are less than 4/100th of an inch (1 mm) in diameter, which makes them impressive predators of many insects.

Predators such as bears, wolves, cats, and wasps rely extensively on odor to locate prey. Snakes detect airborne chemicals in locating prey. It is for this reason that a snake flicks its tongue in and out; after sampling the air, the snake places its forked tongue into an organ located in the roof of its mouth, allowing it to sense remarkably dilute airborne molecules. Some members of the dog family can detect odor-causing molecules as dilute as a few parts per billion. Their characteristically moist noses and large internal nasal surface areas increase the number of particles that are absorbed.

Web-building spiders—predators familiar to most gardeners—rely on their sense of touch to detect prey (plate 127). These spiders generally first become aware of insects entangled in their webs by sensing vibrations through their legs. They then locate their catch more precisely by plucking the radial strands of the web and detecting which vibrations are dampened—in much the same way that a guitar string vibrates differently after a musician shortens it using a finger. Despite their characteristic eight eyes, many web-building spiders have poor eyesight. Spiders that stalk their prey have better vision, and still others use hearing to detect sounds made by distressed prey.

Capturing prey. Capture is probably the least often witnessed phase of predation. Most of our contacts with predators come when we spot a spider on the web, a predatory insect waiting on one of our plants, or perhaps a coyote lurking at the edge of a wood. But when we do witness the capture of prey, we are likely to stop what we're doing to watch. Whether it be hunting bats detected by the light of a back porch bulb, kingbirds (*Tyrannus*) snatching insects in midair, or lizards lunging for an unsuspecting fly, there is something fascinating about watching the organisms around us in the process of obtaining their next meal.

Some hunters use a stalk-and-trap approach, as when a domestic cat slowly and carefully sneaks up on a bird before suddenly pouncing. Other predators pursue prey vigorously. Pursuit can take place over short distances, as when a predatory beetle rushes over a distance of a body length or two, or over long distances, as when a pack of wolves chases down a deer, finally closing in only after the deer is exhausted.

Cooperative hunting is a capture technique with obvious advantages. A gray wolf hunting alone may easily dispatch a small mammal, but it is unlikely to bring down a healthy 1800 pound (820 kg) moose (*Alces alces*). But by hunting in a pack, in which individual wolves share the rewards as well as the effort and risk, even large prey can be subdued. Cooperative hunting is also known in the cat family (Felidae), the hyena family (Hyaenidae), and—of greater relevance in the garden—in some social insects (order Hymenoptera). Weaver ants (*Oecophylla*), for example, are able to subdue much larger ants through cooperation; other ants and wasps cooperatively attack and dismember prey such as caterpillars that are many times their own individual sizes.

Ambush predators sit and wait in place for prey to come to them. Scorpion flies in the genus *Panorpa* hang by their front legs and capture passing flies with their hind legs, which are modified for this purpose. Crab spiders (*Misumena vatia*) sit in flower heads and wait for would-be pollinators or herbivores to arrive. Other ambush predators include web-building spiders and praying mantises. One cost of being an ambush predator is reduced control over what types of prey are captured. A predator that chases down its prey can elect not to pursue a potential meal after it is located. A web-building spider, on the other hand, has limited control over the types of prey that are trapped. Of course, the spider can elect not to wrap or eat a captured organism (and some do release unpalatable prey), but web-building spiders tend to be relatively cosmopolitan in their eating habits and have been found to eat prey from as many as fourteen different orders of insects. But there are also specialists among web builders. A number of orb weavers build nests that have a vertical, nonsticky ladder that can extend 70 inches (180 cm) above the roughly circular catching part of the web. The purpose of this ladder becomes evident when night-flying moths fly into it, then tumble down its length to the spider waiting below.

A remarkable variation on the sit-and-wait theme is seen in bolas spiders (*Mastophora*). Rather than build a sticky web, these spiders dangle a single thread with a drop of adhesive at its end. When an insect flies within range, the spider swings this sticky drop toward it, and if the insect is caught, the spider reels it in and paralyzes it before wrapping and eating it. Some *Mastophora* also secrete a chemical that mimics a sex pheromone produced by female armyworm moths (*Pseudaletia unipuncta*). When a male moth approaches, hoping to find a receptive female, it instead finds a hungry predator. Some predators, such as tiger beetles (family Cicindelidae) and jumping spiders (family Salticidae), use both ambush and active searching to capture prey. Other arthropods use active searching while hunting, but periodically return to a home base to sit and wait for prey.

Predation is usually thought of as an active, even ferocious, struggle. Most predators, however, are more likely to attack prey that are smaller or weaker than themselves, thereby increasing the likelihood of success. Exceptions to this size relationship are seen in predators that use traps (including webs of spiders), those that hunt in groups, or those that inject their prey with venom.

Handling prey. Once its prey has been captured, a predator may still face a daunting challenge. Captured prey may put up a heroic fight, protecting itself with teeth, claws, stingers, and other weapons of defense, which can cause serious harm to a predator. Other species may take a different approach, as when a Virginia opossum (*Didelphis virginiana*) or nine-banded armadillo (*Dasypus novemcinctus*) plays dead. Many predators hesitate to eat prey that is already dead and may abandon an apparently moribund animal—especially if it is armored, protected by spines, or smells bad.

Even after death, prey may cause problems for a hungry predator. Although some predators swallow carcasses whole, others feed selectively to avoid bones, skin, hooves, or parts that are distasteful, injurious, toxic, or of low nutritional value. Birds that feed on winged insects may consume the bodies, but discard the wings—as was probably the case for the luna moth (*Actias luna*) whose beautiful green wings Steve Carroll found in his garden.

Predators that have young to feed may need to carry their prey long distances, either intact or partially digested. And if predators do not wish to become prey themselves, they must be wary of their *own* predators. They may accomplish this by staying alert or by bringing their meal to a safer locale before eating, as when a red-tailed hawk captures a snake in an open field but carries it off to a more protected area before eating it. Black-capped chickadees (*Poecile atricapillus*), red-headed woodpeckers (*Melanerpes erythrocephalus*), and other birds use a similar strategy when they sample from backyard feeders, but bring their "prey" to the relative safety of a nearby tree.

Predators must also sometimes protect their meal from others that would steal it. The thieves and the victims of these thefts may be members of the same species, as in American crows (*Corvus brachyrhynchos*), or they may be members of different species, as when a herring gull (*Larus argentatus*) steals a fish from a smaller laughing gull (*Larus atricilla*). The never-ending battles for carcasses between African lions and hyenas are particularly ferocious.

Defenses Against Predators
The remarkable extent and variety of prey defenses indicate how powerful a force predation is in the natural world. Just as predators can be viewed in terms of how they detect, capture, and handle their food, prey can be grouped according to their abilities to avoid detection, capture, and, ultimately, consumption.

Avoiding detection. In general, the best strategy for staying alive is to avoid being detected by a predator in the first place. Organisms may escape being preyed on by daytime predators by avoiding activity during the day. Many garden caterpillars feed at night and spend the day some distance from the plants on which they feed. This strategy decreases the chances that daytime predators (including gardeners) can use leaf damage or feces to locate them. Many caterpillars also move some distance from their host plants before pupating, which also likely decreases predation.

Many common behaviors probably evolved, at least in part, as a means of minimizing detection. A beetle that travels under the leaf litter may do so to avoid visually hunting birds. Larger animals may take advantage of natural cover provided by shrubs and fallen logs. Burrowing enables animals such as moles to minimize detection by aboveground hunters, although some predators can detect burrowing activity from the surface, and predators such as snakes and weasels (*Mustela*) may enter confined burrow systems in search of a meal. Leaf miners and stem borers (family Cerambycidae), which can be particular problems for gardeners, may minimize predation by feeding inside plant tissue.

Other behaviors help hide the telltale signs of feeding, rather than the perpetrator itself. When herbivores feed in or on leaves, the resulting trails, holes, and ragged edges may alert insectivorous birds looking for a meal, just as they are a signal to gardeners. In one experiment, chickadees learned to associate damaged leaves with the presence of caterpillars (which were secured to these artificially damaged plants by the researchers). This probably explains why some caterpillars hide their damage by snipping off leaves on which they have fed. Some herbivores feed in such a way as to leave behind a symmetrical leaf outline, whereas others actually position their bodies along the damaged leaf edge to fill in for the portion they have eaten. Each of these behaviors decreases the likelihood that the herbivore will be noticed.

Many species minimize prey detection by blending into their environment. Protection through camouflage, known as crypsis, can be accomplished through appropriate colors, patterns, outlines, shapes, and behaviors. Crypsis is particularly common in garden insects. Green grasshoppers and caterpillars feeding on green plant parts have a good chance of being overlooked by their predators—and by gardeners. Fish and birds that are dark above and light below have a pattern of protective coloration called countershading. When viewed from below by a potential predator, a countershaded fish blends into the background of the light sky above; if viewed from above by a hungry gull, its back blends into the darkness of deeper water.

Crypsis can be combined with behavioral strategies such as background selection to minimize detection (plate 128). When resting against the granite bedrock, the canyon tree frog (*Hyla arenicolor*) of the southwestern United States is nearly invisible, whereas against other backgrounds in its environment it is readily apparent. Similarly the gray tree frogs that sat on and called from the peeling, aluminum-painted propane tank in Steve Carroll's yard were much more difficult to see than the one that called from the hose reel (plate 35)—although no experiments comparing predation on these frogs were carried out. Crypsis is especially effective when individuals remain still, as can be appreciated by anyone who has startled or nearly stepped on a well-camouflaged rattlesnake or ruffed grouse (*Bonasa umbellus*). Some species change their appearance to match their environment. Snowshoe hares, for example, change their coloration seasonally, going from a mottled brown in the summer to pure white during the winter months; chameleons (*Chamaeleo*) can change their colors over much shorter time periods.

The underwings (*Catocala*), a group of more than a hundred moth species of the deciduous forest, take advantage of an unusual combination of crypsis and dramatic coloration. When at rest against a tree trunk, these moths fold their drab, bark-resembling forewings over their brightly banded or colored hindwings, making them difficult to see. Laboratory and field experiments have shown that underwing moths tend to select backgrounds that match their wing markings and reflectance patterns. But crypsis does not guarantee protection. If a foraging bird does detect and attack a resting underwing, the moth suddenly uncovers its brightly colored lower wings. It is thought that the sudden appearance of this color startles the bird, which may relax its grip or miss its aim, allowing the moth to escape.

Crypsis is perhaps easiest to appreciate in prey that are hunted visually. But species can also be cryptic to other types of detection. For example, moths in the family Noctuidae (familiar to gardeners as the family that includes cutworms and corn earworms, *Helicoverpa zea* or *Heliothis zea*) evade bat predation by producing sounds at the same frequency as those produced by the bats. The moths are then less easily detected against this background noise.

A particularly interesting form of cryptic defense involves mimicry, in which an organism resembles either a member of a different species or an inanimate object. It is quite difficult to distinguish between a motionless walking stick (family Phasmatidae) and a real stick on the forest floor (plate 129) or between stem-mimicking caterpillars in the family Geometridae and real leaf stems. Behavioral adaptations are also common in species that practice mimicry. Central American mantises in the genus *Acanthops*, which resemble dead leaves, further decrease their chances of being found by hiding their more-visible head under their folded front legs. Mimicry, therefore, is not a case of avoiding detection per se, but of avoiding being identified for what the organism really is—a potential meal. For example, certain swallowtail caterpillars resemble bird feces on the upper surface of leaves. Unfortunately for their victims, predators also practice mimicry. Female fireflies in the genus *Photuris* mimic the light pattern normally produced by females of a different firefly species. When males of the second species arrive to mate, they get eaten instead.

Avoiding capture. Once detected by a predator, all hope for the prey is not lost. Leafhoppers, grasshoppers, and frogs leap away; birds and butterflies take flight; ground squirrels and meerkats (*Suricata suricatta*) dive into their burrows; and fish seem to disappear miraculously before a snorkeler's eyes. Given the choice between fight, flight, or being eaten, many organisms choose flight. Quick escape is particularly interesting in species that move in groups, such as fish schools and bird flocks. Simultaneous movement by so many individuals seems to confuse predators—something that you can test by trying to follow the movements of a particular fish or bird the next time you find yourself snorkeling or walking across an open field.

Although attempting to escape is common in nature, there are also many species whose members stand their ground. Turtles, armadillos, hard-shelled insects, and others rely on their protective armor. Some retreat completely into their shells, where they are out of reach or at least better defended against predators. Other prey animals use posturing, in which the potential prey attempts to make itself appear larger, more troublesome, or less desirable than it really is. A lizard may extend its collar, a bird may puff out its feathers, and many species of fish swell up like balloons, all of which may deter a predator or buy time for escape.

One of nature's most unusual defenses is the ability to sacrifice parts. Some butterflies have wing patterns that cause birds to strike the less essential hind wings, rather than the head or body. Hairstreak butterflies (family Lycaenidae) with tattered hind wings are common, a result of attacks aimed at the false-head patterns on their wings. The so-called glass snakes (actually lizards, *Ophisaurus*) rather easily lose their tails, which distract pursuers by continuing to wriggle. Regeneration of lost parts is sometimes possible, as when a salamander in the garden replaces its tail or a leg. (Of course, from the perspective of a gardener, this capability should not seem all that odd, given how well plants do the same thing!)

Many social burrowing mammals, including North American ground squirrels and woodchucks (also called groundhogs, *Marmota monax*), give out warning calls when a predator is spotted. This call often precedes a dash to the safety of the burrow and warns relatives to take cover as well. Alarms can also serve as a call-to-arms; crows give an alarm when an owl or hawk is spotted, causing other nearby birds to mob the larger intruder. The flashing of their white rump patches by pronghorn antelope (*Antilocarpa americana*) serves a similar purpose.

The natural world encompasses an amazing variety of chemical defenses. In one survey of antipredator mechanisms in arthropods (including insects, spiders, and their allies), 46 percent were found to use some form of chemical defense. Some social insects produce chemical warnings when they are injured; these warnings, when detected by other group members, enable them to mount a defensive or counter-offensive response. This type of signal-response combination is common in bees, hornets, ants, and termites. Chemical communication in aphids does not lead to aggression, which would likely be of little value against their larger and stronger predators; instead, it causes the aphids to drop to the ground. Aphids use this defense against foraging beetles as well as humans attempting to pick them off garden plants.

When bombardier beetles (*Brachinus*) are threatened, they mix together a variety of reactants and enzymes in special abdominal reaction chambers. When these chemicals combine, a hot toxic mixture is expelled in the direction of the attacker. In the ocean, cuttlefish (order Sepiida) and octopuses (family Octopodidae) are known for their well-directed blasts of ink that screen their escape. By the time this screen dissipates enough for a predator to see, the responsible party is long gone. Although you are not likely to meet an octopus in the garden, you—or more likely your dog— may well find skunks (*Mephitis* and *Spilogale*, for example) with their well-known and odorous chemical defense.

Less dramatic chemical defenses may still be effective. A millipede can discourage the advances of a predator by curling up into a tight ball and secreting a noxious chemical through pores in its skin. Many gardeners have been "slimed" by millipedes that they've picked up. Daddy-longlegs secrete a large variety of defensive chemicals from special glands, and termites combine chemical defense with physical attack. Still other arthropods regurgitate when threatened, as when a grasshopper spits into the palm of a curious child's hand. Even more unusual is a tactic used by some beetle larvae, which wear their own feces as a deterrent or shield against predators.

Some organisms avoid attack by virtue of their reputations. South American poison-dart frogs (*Phyllobates terribilis*), which produce potent neurotoxins that deter predators from feeding, advertise themselves with bright coloration. One of the best known animals in this group is the monarch butterfly, which most predators avoid because of its unpleasant taste. As mentioned, the monarch obtains its chemical defense, cardiac glycosides, from milkweed plants consumed as a caterpillar. But, just because an organism produces a toxic compound does not free it from predation—as natural selection has also acted on predators' abilities to counter these defenses. As distasteful as monarchs are, black-headed grosbeaks (*Pheucticus melanocephalus*) and a few other bird species eat them in huge numbers on their overwintering grounds.

Animals use mimicry to avoid capture as well as to go unnoticed. In some cases, a palatable or innocuous species mimics a toxic, distasteful, or dangerous species; this is known as Batesian mimicry after the biologist who first described it. Examples include *Spilomyia fusca*, a syrphid fly that mimics hornets, and nonpoisonous false coral snakes (*Pliocercus*) that mimic the venomous coral snakes (*Micrurus*). This kind of mimicry can also occur at the chemical level. Certain rove beetles (*Creophilus maxillosus*) not only look and sound like bees and wasps, they also smell like them. Many North American gardeners know about the viceroy butterfly's

(*Limenitis archippus*) mimicry of the distasteful monarch butterfly (plate 130). This example has long been thought of as the classic example of Batesian mimicry, but recent research has shown that the viceroy can also be distasteful, making this relationship more complicated than once thought. (As a further interesting complication, in regions where monarch butterflies are rare, many viceroys are reported to more closely resemble the monarch's relative, the queen butterfly, *Danaus gilippus*, instead.)

The effectiveness of Batesian mimicry does have its limitations. As long as the distasteful model species is more common than the palatable mimic species—that is, if predators are more likely to encounter it as they feed—subsequent feeding on *either* species will be discouraged. If the mimic becomes too common, however, predators will encounter palatable mimics more and more frequently. They will then continue feeding on this model-mimic pair, leading to at least a temporary decline in both species. As a result, the mimic is generally prevented from becoming too common.

A different form of mimicry, known as Müllerian mimicry, occurs when *all* of the species involved are distasteful or dangerous in some way. Their common appearances and defense mechanisms tend to reinforce the message that predators should look elsewhere for a meal. The common yellow-and-black banding pattern of many stinging bees and wasps is generally interpreted as an example of this type of mimicry. In New Guinea, several birds in the genus *Pitohui* have colored hoods, the skin and feathers of which contain a toxic alkaloid, and this too is interpreted as Müllerian mimicry. The black-and-white patterning of skunks is an example more likely to be encountered in North American gardens.

There are many well-known species, from coral snakes to skunks, that exhibit warning markings, or aposematic coloration. An explanation for aposematic coloration was suggested by Alfred Russel Wallace, naturalist, collector, and codeveloper of the theory of evolution by natural selection. Charles Darwin wrote Wallace to ask his thoughts on such color patterns. In response, Wallace reasoned that "distastefulness alone would be insufficient to protect a larva unless there were some outward sign to indicate to its would-be destroyer that his contemplated prey would prove a disgusting morsel, and so deter him from attack." In effect, then, aposematic coloration is negative advertising—and it is these aposematically colored species that are the models in Batesian mimicry.

The effects of predation can also be combated through the power of large numbers. Some organisms produce enormous numbers of young within short periods of time, thus outstripping potential predators' ability to consume them all (plate 131). Others congregate in large groups and thus decrease any particular individual's odds of becoming a predator's next meal. The masting of seed and fruit crops is an example of such safety in numbers. Another example involves consumption of periodical cicadas (*Magicicada*) by insectivorous birds. Ecologists have found that when adult cicadas first emerge, 15–40 percent are consumed by birds. However, as emergence peaks, a smaller and smaller percentage of cicadas is eaten. Anyone who has witnessed a large cicada emergence and has listened to their deafening roar over the ensuing weeks can appreciate the fact that many, many cicadas manage to avoid being eaten.

What if prey are captured? For individuals of most species, once captured, the probability of injury or death is greatly increased. However, as noted in the section on predators' handling of their prey, a captured meal is not necessarily an eaten meal. Prey may escape from or be released by their captors, often thanks to teeth, claws, stingers, or unpleasant chemicals. Even if these structures are used primarily in feeding, they can also be effective in defense. For instance, even normally nonaggressive mice have sharp teeth, rabbits have powerful kicking legs, snakes have fangs (and sometimes venom), and so on. Any gardener who has been stung by a bee, bitten by a snake, or sprayed by a skunk knows well the power of such defenses. Although we humans generally attack these creatures accidentally, a hungry predator would receive similar treatment while trying to turn one into a meal.

Parasitism

Parasitism is similar in many ways to herbivory and predation because one organism (the parasite) benefits at the expense of another (the host): a win-lose relationship. But there are significant differences in how the beneficiaries extract resources from their victims. Parasites generally associate intimately with their hosts for extended periods, often even living within their tissues, whereas most predators or herbivores eat their prey from the outside and move away when satiated. Parasites tend to be more specialized in their adaptations to their hosts than are predators or herbivores to their prey, probably due to the need for extended coexistence. Because parasites need their hosts to continue living, their injury to hosts tends to be subtle and gradual—often manifesting itself by progressive debilitation, in contrast to the usually obvious and immediate physical trauma resulting from predation and many cases of herbivory. These differences in offensive tactics necessitate different defensive measures; therefore, it is useful to distinguish parasitism from herbivory and predation. It is also good to recognize that there are borderline cases not clearly falling into either category.

Relationships between parasites and their hosts range widely. Toward one end of the spectrum are facultative parasites that can also survive as free-living detritivores. These parasites may do such severe damage that hosts die or slough off a lot of dead tissue. Some of the most severe ailments of garden plants are caused by facultative parasites, including damping-off caused by *Pythium* fungi and various conditions caused by the fungus *Rhizoctonia solani*. At the other end of the spectrum are obligate parasites, which are so specifically adapted to conditions within their living hosts that they cannot survive elsewhere (plate 132). Obligate parasites sometimes do little overt damage to their hosts. A number of root-dwelling bacteria and downy mildew fungi fall in this category.

Interactions of Parasites with Their Hosts

Parasitic relationships involve several necessary components and stages. First, a parasite must survive somewhere (a reservoir) until it encounters a suitable host. A parasite may persist as an inhabitant of another living host, especially in the case of obligate parasites; as a dormant spore or cyst or other metabolically inactive form; or even as a free-living organism, as in the case of many facultative parasites.

Next, there must exist an accessible and susceptible host. If a potential host exists but it is not in a susceptible state, no infection by the parasite can arise. It is well known in organic gardening practices that plants overfertilized with nitrogen are often more susceptible to fungal infections than are more moderately nourished plants, as they tend to have tender, watery tissues. Low light levels can also increase susceptibility of plants to many pathogenic parasites due to a reduced ability to synthesize defensive materials.

Some obligate parasites require different hosts at different stages of their life cycles. In such cases, if even one host is absent, the parasite cannot complete its life cycle and will eventually die out. Examples of plant parasites requiring multiple hosts include *Puccinia* rust fungi of wheat and other grasses, whose alternate host is barberries (*Berberis*); the white pine blister rust fungus *Cronartium ribicola*, whose alternate hosts are currants, especially black currants (*Ribes nigrum*); and the cedar-apple rust fungus *Gymnosporangium juniperi-virginianae*, which affects cedars and junipers (*Juniperus*) and whose alternate hosts are pome fruits such as hawthorns (*Crataegus*) and apples (plate 133). In the 1930s, the requirements of these destructive parasites for alternate hosts led to U.S. government–sponsored programs to exterminate the less economically valuable hosts—barberries, currants, and eastern red cedars (*Juniperus virginiana*), respectively. It is still illegal in some parts of North America to raise black currants because of the threat they pose to valuable conifer forests by hosting white pine blister rust.

If contact between a potential host—no matter how susceptible—and a parasite does not occur, there will be no infection. An agent that transports a potential parasite from its reservoir to a potential host is known as a vector. Wind, insects, rain drops, or a gardener's fingers, tools, or shoes can act as vectors. Quarantines—whether established by posting customs officials at a border crossing, excluding pests with a fine-meshed gauze, or growing one's own transplants from seed so as to avoid introducing soil from other gardens—are attempts to interdict vectors and prevent contact between potential parasites and prospective plant hosts.

Once contact between a host and parasite occurs, the parasite must be able to recognize the host, overcome or evade its defenses, and establish itself in or on the host—that is, the parasite must be able to infect the host. Establishing an infection is often a complex process and may involve a number of reciprocal actions and reactions on the parts of the participants. It is appropriate at this point to distinguish between *infection* and *disease*, terms often confused and misused. Infection is the establishment of a parasite (or symbiont, see the next section) within a host and its subsequent growth or reproduction. Disease is disruption of an organism's metabolism such that it cannot function satisfactorily. Infection does not always lead to disease, and disease is not always the result of an infection.

Some disease-resistant varieties of garden plants may have altered cell-wall chemical markers and thus escape being parasitized through a form of chemical camouflage, as parasites need to detect particular carbohydrate or protein molecules on the cell walls of prospective hosts to initiate invasion. Absence of the proper molecules will result in no invasion. Thus, a minor change in cell surface chemistry can be a powerful thing! In other cases, plants have slippery or hairy surfaces to which potential parasites may not be able to adhere. Many disease-resistant varieties of tomatoes, potatoes, and other plants are festooned with dense mats of tiny hairs that present formidable barriers to some parasites (plate 134).

Once a parasite successfully adheres to a potential host and recognizes that it "has arrived," it may still face obstacles to physically entering the host. Many plants have hard or rubbery waxes, woody material (lignin), silica or other minerals, or even natural, plasticlike polyesters on their surfaces or embedded in their cell walls. Cell walls themselves can be thick, reinforced blends of tough fibers and rubbery gums or gels, especially in mature plant tissues. Thus, many parasites can enter hosts only through weaker, immature tissues or via traumatic injuries such as abrasions, tears, or insect feeding punctures. Some parasites hitch a ride in the gut or mouthparts of an insect, nematode, or other invertebrate and enter a susceptible host plant when the vector organism feeds.

Viruses face a special challenge in invading plants because they have no means of their own for penetrating plant tissues. Most viruses that infect plants are wholly dependent on animal vectors to transport them from host to host and to introduce them into host tissues. Some bacteria can enter plant or animal tissues by secreting digestive enzymes that dissolve the intercellular glues that bind cells together (causing the tissues to fall apart), but many depend on traumatic injury for introduction. In contrast, although trauma may facilitate the entry of fungi into hosts, in general they are not as dependent on injuries as are viruses and bacteria. Some parasitic fungi enter plants via stomates; others are capable of entering hosts by weakening their armor using enzymatic digestion and then penetrating with hydrostatic pressure-driven haustoria (stubby projections from fungal mycelia).

Even if a parasite enters a potential host, it may not succeed in establishing itself unless it overcomes an array of internal defenses. Plants apparently lack immune systems of the type found in animals (although they have a definite, yet poorly understood, immunelike response to repeated pathogen exposure), but they have at their disposal a wide variety of biochemical defenses with which they can restrict, debilitate, or even kill invading parasites. For example, many plants continuously synthesize and release compounds highly toxic to other organisms or store such compounds in glands ready for immediate release. Also, a great many plants quickly synthesize toxic substances (phytoalexins) in nonspecific response to trauma such as herbivory or invasion by parasites. One version of this defensive chemical strategy is the hypersensitive response, in which plant cells in the vicinity of an invading parasite saturate themselves with nonselective poisons and then die, thus creating a toxic zone of death that surrounds the invader and may prevent it from penetrating more deeply. (Humans did not invent pesticides.) But many garden plants, especially those that are edible, have been selected by humans over the millennia specifically to be nontoxic. As a result, they are often more vulnerable to herbivores and parasites than are their wild relatives. (It might be questioned whether efforts to decrease use of synthetic pesticides by breeding disease resistance back into edible plants, including the ability to make natural toxins, necessarily make them safer for human consumption.)

Plants also use other defensive measures. Many plants, especially legumes, release carbohydrate-binding proteins (lectins) that interfere with microbial cell division and growth. Others remove and sequester soluble nutrients in parts of the plant beyond the immediate reach of invaders. Yet others release digestive enzymes that selectively attack bacterial or fungal (but not plant) cell walls; and some plants physically slough off invaded parts, thus saving the rest of their tissues.

Of course, parasites have developed their own tactics and weaponry to overcome host plant defenses. Some parasites do little initial injury and conceal or lack biochemical markers that may signal their presence—a sort of stealth approach, so that plant hosts do not mount an active defense until it's too late. Apparently, some well-adapted obligate parasites are hardly even sensed by their plant hosts; most damage to hosts results from a debilitating drain of nutrients, not from overt injury. Other parasites develop resistance to plants' defensive toxins, release enzymes to digest plant walls and other physical barriers, or even (in the case of some viruses) wage genetic warfare by taking over a host cell's genetic machinery and turning it against the plant itself. Some parasites, especially some facultative parasites that can survive even if their host dies, use brute force, releasing potent poisons and aggressive enzymes that do massive damage and reduce plants' abilities to resist further invasion.

Even if a parasite successfully establishes itself in a host, it will ultimately be a biological failure unless it reproduces and disperses its progeny to colonize new hosts. Most viruses and many bacteria multiply within host plant tissues and are then transmitted to new hosts by feeding insects, especially aphids and leafhoppers. Some viruses integrate themselves so deeply into the cells of the host that they are incorporated into new seeds, and a young plant may actually be born infected. The plum necrosis virus is a notorious example of such a maternally transmitted virus (plate 135). This virus has so far stymied efforts to introduce otherwise highly desirable central Asian stone fruit (*Amygdalus* and *Prunus*) varieties into North America. Before commercial distribution, seeds of cabbage and other brassicas are often immersed in hot water (140°F, 60°C) for several minutes to reduce seed-borne viral diseases such as cabbage yellows. Bacteria such as *Pseudomonas solanacearum* (now renamed *Ralstonia solanacearum* but still widely known by its historic name) and *Erwinia atroseptica* (recently renamed *Pectobacterium carotovorum* ssp. *atrosepticum*) and many pathogenic fungi can be similarly transmitted to future plant generations by colonization of seed coats. However, most fungi and some bacteria are transmitted to new hosts by wind-blown or water-splashed spores released from parasites' fruiting structures. Such spores may persist in the environment for quite some time, awaiting favorable conditions for germination and colonization of a new host. In an extreme case, dormant vegetative structures of *Fusarium* and *Verticillium* fungi may survive for many decades in the soil, making it almost impossible to eradicate these parasites by using crop rotation (plate 136).

Microbial Parasites of Plants

Because they are generally well adapted to the acidic, carbon-rich, nitrogen-poor, highly aerobic environment that characterizes most plant tissues, fungi cause the great majority of plant diseases, especially under humid conditions (plate 137). Ascomycetes and imperfect fungi are responsible for most fungal diseases of plants, although significant parasitic species exist in all groups of fungi. Some of the more prominent ascomycetous fungal parasites of North American garden plants include *Monilinia* species, which cause brown rots of stone fruits such as cherries and peaches; *Ceratocystis ulmi*, which causes Dutch elm disease; *Gibberella* species, which cause diverse blights of grasses; *Sclerotinia sclerotiorum*, which causes watery soft rot of many vegetables and flowers both in the garden and after harvest; and *Venturia inaequalis*, which causes scab of apples. Widespread imperfect fungal parasites include soil-dwelling *Fusarium* and *Verticillium*, which cause wilt diseases of a wide range of food and ornamental plants. Others include *Alternaria*, *Colletotrichum*, and *Helminthosporium* (also known as *Drechslera* and *Bipolaris*) that spread by means of air-borne spores and cause foliar diseases of diverse plants. The notoriously ferocious plant pathogen *Rhizoctonia solani* is also an imperfect fungus. It causes damping-off of innumerable seedlings under cool or moist conditions and can devastate virtually every plant species—and even attacks other fungi.

Basidiomycetous fungal parasites are best known for causing diseases of grasses and woody perennial plants. Species that cause smuts, bunts, and rusts of grasses (including important cereal grains) are all obligate parasites, although most require alternate hosts to complete their life cycles. Blister rust of pine trees is caused by the basidiomycete *Cronartium ribicola*. Most decay of living woody trees—heart rot, leading to hollow trees, breakage of trunks and limbs, and death of roots—is caused by the basidiomycetes *Armillaria mellea* (synonym: *Armillariella mellea*) or *Fomes* species (plate 138).

Major parasitic oomycetous fungi include members of the genus *Phytophthora* (literally, "plant destroyer"), which cause blights and wilts of diverse plants including avocados, raspberries, and citrus fruits as well as sudden oak death. (*Phytophthora infestans* devastated the Irish potato crop in 1846–1848.) *Pythium* species cause damping-off of emerging seedlings. *Peronospora* species and their relatives cause a variety of downy mildews, including blue mold of tobacco and related plants. Two major plant diseases caused by plasmodiophoromycete fungi are clubroot of cabbage and other brassicas and powdery scab of potatoes.

It is fortunate that there are few actinomycete pathogens of plants, because there are few effective controls for those that do exist. Some actinomycetous pathogens include *Streptomyces scabies* and relatives that cause scab of potatoes and some *Corynebacterium* species that cause bacterial canker of tomatoes and other plant diseases. Limited control of actinomycetes may be achieved by environmental manipulation, such as by adding sulfur to soil to reduce the pH below 5.

In general, bacteria are less well adapted to live in plant tissues than are fungi. Nevertheless, there are numerous bacterial diseases of plants, many of which are fatal. (In contrast, most viral and fungal ailments may debilitate and disfigure but do not often kill outright.) Bright yellow *Xanthomonas* bacteria cause diseases such as black rot of cole plants, blight of beans and other legumes, and leaf spot of mallows. *Erwinia* species cause the deadly wilt of cucumbers and melons (*Cucumis melo*; plate 139); fire blight of pears (*Pyrus*) and apples and other rose family members; and soft rot of numerous plants. *Agrobacterium* species cause tumorlike galls. *Pseudomonas* bacteria cause diverse leaf and fruit spots.

However, mycoplasmas—amoeboid bacteria that lack cell walls and that are abundant in garden soils rich in incompletely decomposed organic matter—may be the most widespread and devastating bacterial plant pathogens. In recent years mycoplasmas and mycoplasma-like organisms (MLOs) have been found to cause more than forty plant diseases known as yellows, stunts, and declines, including deadly aster yellows (the leading disease of asters and related plants), mulberry dwarf disease, and alfalfa mosaic (caused jointly by an MLO and a virus).

Insect-transmitted viruses cause destructive diseases such as mosaics of tobacco, tomato, turnips, cucurbits, and many other plants; curly top of beets and relatives; potato virus X disease; many tumors and galls; and various streak diseases (plate 140). Plant viruses rarely kill their hosts outright, but tend to physiologically debilitate them, thereby weakening them in the face of environmental stress or nonviral pathogens; or they may cause reproductive sterility. Viruses also cause variegation, mottling, streaking, twisting, or curling of stems and leaves, effects that are actually exploited by some ornamental plant breeders to produce multicolored or ornamentally distorted foliage or flowers.

So ubiquitous are viral infections among vegetatively propagated perennials such as fruit trees, woody ornamentals, and plants raised from tubers or bulbs, that infected conditions have long been accepted as normal. In fact, when such plants are cured of viruses, either accidentally or by application of modern technology, such as meristem culture of brambles (*Rubus*), the resulting gains in vigor are often phenomenal. The economics of producing and planting virus-free perennial plants may be questionable, though, because insects such as aphids and leafhoppers usually ensure that outdoor-grown plants do not remain virus-free for long. Nevertheless, virus indexing (determining the viral infection status of plants and culling out infected plants) is now standard procedure in many nurseries producing plants for commercial orchards and landscapers.

Microbial Parasites of Microbes and Garden Animals

Bacteria can parasitize fungi, and viruses can parasitize fungi and bacteria. Indeed, viral diseases seem to be among the more effective biological controls of bacterial and fungal populations in nature, although their significance in garden environments has been little studied. Some fungi are hyperparasites; that is, they parasitize plant-parasitic fungi, thus possibly aiding in control of diseases of plants caused by fungi.

Many fungi selectively parasitize insects, reducing their mobility, inhibiting their feeding, and often killing them. In fact, parasitic fungi and bacteria are major controls of insect populations, possibly more significant than predators and pesticides. Fungi are also one of the major natural controls of nematodes, amoebas, rotifers, and other soil-dwelling microfauna. Fungi of the order Zoopagales are especially interesting: they snare their prey with touch-sensitive nooses formed from fungal filaments or with sticky, knobby projections and then parasitize the immobilized animals.

Animal Parasites of Plants

Many nematode-plant relationships are considered parasitic, but tend toward the predatory end of the spectrum. Many ectoparasitic nematodes—those that do not enter the tissues of their hosts—feed on a host plant until it is debilitated or dead or until the plant successfully repels them by secreting toxins. The nematodes then move on in search of a fresh victim—much as do herbivores. Other ectoparasitic nematodes may spend most of their life cycle firmly attached to a single host plant; only their newborn juveniles are motile, roaming until they encounter a suitable host or die. Endoparasitic nematodes embed themselves partially or completely within host tissues and feed from the inside. Nematode-like nematomorphs are truly endoparasitic, dwelling entirely within host plant tissues, and absorbing nutrients from their hosts through their thin, permeable skins. The eggs or juveniles of endoparasitic nematodes and nematomorphs are released only upon breakdown of the host plant tissues.

Plants can resist nematodes and nematomorphs in various ways. Some plants have waxy, woody, or silica-impregnated surface layers impenetrable to the stylets of nematodes. Others secrete toxic substances either continuously or in response to nematode invasion. Still others use a hyper-sensitive response as described in plant defenses to microbial parasites.

Some tiny wasps and mites are parasitic on higher plants, especially perennials of the rose, aster, and oak families. These insects lay their eggs under the epidermis of stems or leaves; the eggs then hatch into larvae that feed inside the plant tissues. Although the amount of plant tissue consumed by these insects is usually of minor consequence, many of the larvae produce or induce the production of plant growth hormones, causing galls to grow (plate 141).

Animal Parasites of Garden Animals

Many wasps lay their eggs on or in the larvae of insects (plate 142). When the eggs hatch, the wasp larvae burrow into and through the bodies of the host larvae, gradually eating them alive from the inside out. Parasitized larvae become sluggish, cease to feed on plants, and eventually die. The mature larvae of the parasites then undergo metamorphosis to release new adult wasps, which fly off, mate, and seek fresh victims. This parasitism—a sort of slow-motion predation called parasitoidism—is significant in controlling populations of many insects such as moths, butterflies, flies, mosquitoes, aphids, whiteflies, and even cockroaches. Thus, gardeners would be well advised to leave parasitized caterpillars alone so that they will yield another generation of parasitoid wasps. Adult parasitoid wasps are strictly herbivorous, feeding primarily on floral nectars of plants of the umbel or celery family. Some gardeners plant flowering umbelliferous plants, such as dill (*Anethum graveolens*), fennel (*Foeniculum vulgare*), caraway (*Carum carvi*), or Queen Anne's lace, in or near their gardens specifically to attract and sustain beneficial parasitoid wasps (which do not sting humans). Wasps and bees face their own parasites, of course. For example, parasitic *Varroa* mites and tracheal mites (*Acarapis woodi*) have severely reduced populations of honey bees in North America.

Tachinid flies similarly spend their larval stages as internal parasites of insects, spiders, centipedes, and other arthropods. Tachinid flies are not as well known as parasitic wasps but may be even more ecologically significant because they tend to be less picky about their hosts. Tachinids parasitize larvae of insects that undergo complete metamorphosis as well as nymphs of insects that undergo incomplete metamorphosis and even adult stages of some insects such as beetles and grasshoppers. One genus of tachinid flies is very undesirable from the viewpoint of humans, however: *Sturmia* species are serious pests of the Oriental silkworm (*Bombyx mori*)—although the white mulberry trees (*Morus alba*) that are silkworms' main dietary staple may appreciate its efforts!

Parasites of mammals such as hookworm of dogs (*Ancylostoma caninum*) and roundworms of cats and dogs (*Toxocara cati* and *T. canis*, respectively) may be picked up by contact with garden soil contaminated with feces of host animals (or ingestion of plants contaminated with such soil). Parasitic dog fleas (*Ctenocephalides canis*) and avian fleas (*Echidnophaga gallinacea*) pass through a maggot-like larval stage that may feed on decaying plant materials in gardens; after metamorphosis the adults hitch a ride on passing animals—or gardeners.

Plant Parasites of Plants

Mistletoe (*Phoradendron* in North America; *Viscum* in Europe) may have a romantic reputation around Christmas, but it is also a parasite of woody perennial hosts in mild and warm climates. Dwarf mistletoe (*Arceuthobium*) is a major pest of conifers in western North America. Other common parasitic plants include dodder and numerous orchids (family Orchidaceae), including the aromatic vanilla plant (*Vanilla planifolia*). (Although vanilla may also be raised as a free-living plant.) To varying degrees, these parasitic plants have lost the ability to extract water and other nutrients from the soil; they therefore must attach themselves to and send probing rootlets into the veins of host plants. Damage to hosts may range from imperceptible in cases of light infestations by orchids or mistletoe to severe in cases of heavy dodder infestations of cereal grains. Dodder has largely lost the ability to photosynthesize, and thus it must depend on host plants not only for water and mineral nutrients, but for photosynthesized carbohydrates as well. One might question how dodder could even be classified as a plant in view of its inability to photosynthesize. The classification is made on the basis of structures such as vascular tissues, flowers, and seeds, which clearly place it in the plant kingdom and distinguish it from other nonphotosynthetic organisms (plate 143).

Parasitism as a Way of Life

Parasitism is ubiquitous and important in the garden as well as in nature. Despite the great costs to parasites of overcoming host defenses, the benefits are probably also considerable, judging by the large numbers of organisms that have adopted this lifestyle. The result of the unending see-saw battle between plants and their parasitic neighbors is that, in nature, few mature plants are free of parasitic infection, yet many are able to grow and survive long enough to produce offspring. The problem for gardeners is that most don't simply want their plants to survive and produce some seed; they want them to grow big, beautiful, and productive—and thus are unwilling to tolerate the condition of constant, moderate sickness that is often the norm for plants in nature. For this reason, many gardeners employ pesticides, quarantines, soil sterilization, manipulation of nutrients and light levels, and other tools and tactics to try to keep plants in an unnaturally healthy state. In chapter 6 we discuss some ecologically sound strategies for managing garden parasites as well as the many other organisms inhabiting our gardens.

Mutualism

Mutualism is a long-term, interactive association involving two or more species in which all participants benefit. In general, mutualisms are win-win situations, although all members may not share equally in the benefits at all times or under all circumstances. In some cases, one member of a mutualism may temporarily exploit or become separated from the other. Thus, a mutualism can be thought of as mutual exploitation that happens to provide net benefits for all participants so that, in the long run, the associated organisms are better off together than they would be apart. Benefits to garden residents involved in mutualisms include shelter from harmful environmental factors; protection from predators, parasites, or competitors; access to otherwise unavailable nutrients or water; enhanced reproduction or dispersal of offspring; and changes in internal metabolism that increase fitness. (Gardeners should be aware that in many publications the word *symbiosis* is used erroneously as a synonym for *mutualism*. Properly speaking, symbiosis is a broader term that encompasses a range of interactive relationships between organisms, including cases of mutualism, commensalism, and even parasitism.)

Commensalism is an intimate, long-term association between two or more species in which one benefits but the others receive neither net benefit nor harm—in other words, a win–no effect relationship. For example, many organisms obtain specific nutrients from waste substances generated by others and cannot survive in the absence of their benefactors, yet they provide no known reciprocal benefit or harm to those benefactors. This type of relationship is both common and important in garden ecosystems. Commensalism is crucial to the functioning of many nutrient cycles, which are discussed in chapter 4.

Plants and Nitrogen-fixing Microbes

One of the best-studied mutualisms is that between *Rhizobium* bacteria and legumes, as well as a few members of other plant families, notably the wild olive or oleaster family (Elaeagnaceae) and the buckthorn family (Rhamnaceae). In the absence of hosts, *Rhizobium* bacteria are free-living and are a minor part of the bacterial microflora of many soils.

As a legume or other compatible plant germinates and grows in soil containing rhizobia (*Rhizobium* bacteria), the young roots secrete chemicals that are attractive to nearby bacteria. If a *Rhizobium* bacterium is of the proper genetic type, carbohydrates on the surfaces of the bacterial cells will bind to proteins (lectins) on the plant root surfaces, which glue the bacteria to the plant root. Because the carbohydrates vary from one strain of rhizobia to another, only particular rhizobia can bind to a given species of plant. For example, rhizobia that can bind to garden peas cannot bind to (inoculate) soybeans (*Glycine max*), true clovers, or alfalfa. Thus, if a gardener wishes to benefit from introducing rhizobia into the soil, it is important to use the strain appropriate to the plants being raised.

If binding between rhizobia and the plant root is successful, the microbes will attempt to digest through cell walls into root hair cells. The plant may react as if it were being invaded by a parasite and attempt to destroy or expel the rhizobia using toxins or digestive enzymes—and sometimes it succeeds. However, if the strain of rhizobia is sufficiently compatible with the host plant, the bacteria somehow overcome or suppress these plant defenses and induce plant cells to form infection threads—long, narrow tubules that aid the movement of the now-multiplying rhizobia into the plant tissues.

Once the rhizobia have taken up residence within root cells, they change form dramatically, becoming swollen, amoeboid cells (known as bacteroids) incapable of survival outside of the host's tissues. Biochemical changes also take place in both the plant and the bacteria, and the two organisms begin a life of cooperative metabolism. For instance, the molecule leghemoglobin, made by neither organism alone yet critical for nitrogen

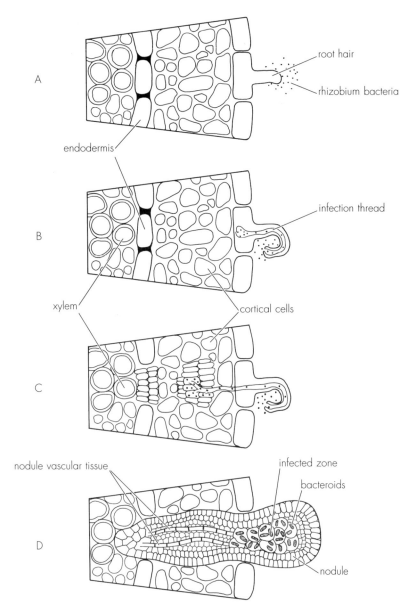

Figure 9. Infection of plant roots by *Rhizobium* proceeds through a sequence of steps involving both the bacteria and host plants. (A) Rhizobia in the soil bind to compatible plant roots. (B) Bacteria digest their way into plant tissues and stimulate plant cells to form an infection thread. (C) Once deep inside plant tissues, rhizobia differentiate into amoeboid bacteroids. (D) Plant cells then multiply into a tumorlike nodule, and the enclosed bacteroids carry out nitrogen fixation.

fixation by the bacteroids, is synthesized with parts of the molecule made by the plant and parts by the bacteroids. The bacteroids also stimulate the production of plant growth hormones, causing surrounding plant cells to divide and grow into a tumorlike mass known as a nodule, which becomes a home for the multiplying colony of bacteroids (plate 144). The rhizobia, which are weak competitors in the open soil environment, flourish in the root nodules, obtaining shelter, water, carbohydrates, and mineral nutrients from their host plants.

So far, this mutualism resembles parasitism by gall-forming bacteria, with most of the benefits accruing to the bacteria and most of the costs being borne by the plant. In some cases, the relationship develops no further and "ineffective nodulation" is said to have occurred—the plants have simply been infected by a parasite that will be sheltered in its roots (or in stem nodules in the cases of some tropical plants), and the bacteria will proceed to steal carbohydrates, water, and other nutrients. However, in most cases "effective nodulation" occurs, and rhizobial infections become mutualisms as the nodule-enclosed bacteroids proceed to convert inert atmospheric nitrogen into ammonia and amino acids that will nourish both the rhizobia and their host plant.

The *Rhizobium*-plant mutualism often enables legumes and other compatible plants to survive or even thrive in nitrogen-poor soils that may support few other plants. Vast quantities of inert atmospheric nitrogen are fixed annually by members of this symbiosis, and it is critical for life on Earth as we know it. Despite these benefits, however, plants may continue to react to colonies of bacteroids as if they were parasitic invaders. Most nodules eventually die from accumulation of plant toxins and are sloughed off into the soil. During breakdown of the sloughed-off nodules some bacteroids revert to free-living forms and escape into the soil so that they can infect other suitable plant hosts, and the life cycle of the mutualism begins anew. Digestion of dead nodule tissues by other soil microbes also releases accumulated nitrogen-containing compounds, so that many other organisms benefit indirectly from the *Rhizobium*-plant mutualism.

A mutualism between alders and *Frankia* actinomycetes also involves nodule formation and nitrogen fixation. In other mutualisms between plants and nitrogen-fixing microbes, the microbes do not invade plant roots, but usually live in close proximity to them, feeding off of nutrients leaked from the roots. These noninvasive mutualisms include the relationship between various tropical grasses and root zone–dwelling *Azospirillum* bacteria and that between the blue-green alga *Anabaena azollae* and the aquatic fern *Azolla*, which has maintained the nitrogen fertility of many Asian rice paddies for thousands of years.

Mycorrhizae

Less well known than rhizobia, but of comparable significance to plants, especially perennials, are mycorrhizae (meaning "fungus-roots"), which form when particular fungi infect the roots of plants. These fungi take photosynthesized carbohydrates from their plant hosts, but in exchange they provide significant benefits, such as water and mineral nutrients (especially phosphorus, zinc, copper, and iron) that the fungi are far more efficient at extracting from the soil than are plant roots. The fungi also provide protection from soil-dwelling pathogenic parasites. In fact, mycorrhizae are so crucial to the survival of many plants that some biologists believe that their development was a prerequisite to the establishment of terrestrial plant life.

There is more than one distinct type of mycorrhizae. Ectomycorrhizae are formed when roots of woody plants, especially certain conifers, are completely encased by a thick sheath of fungal filaments (hyphae). The roots become stubby and highly branched, but a webbed network of hyphae may extend for many feet in every direction from the colonized roots. These hyphae extract water and nutrients from the soil and transport them to the mycorrhizal roots with great efficiency. The fungal filaments usually penetrate only a little way into the roots, often only a few cell layers deep, and normally between cells, not into them. Many mushrooms, including truffles (*Tuber*), boletes (family Boletaceae), and morels (family Morchellaceae), are the reproductive structures of ectomycorrhizal fungi. The mycorrhizal nature of many mushroom species is one reason why they are very difficult, if not impossible, to raise in artificial culture.

Endomycorrhizae (also widely called vesicular-arbuscular mycorrhizae [VAM] or arbuscular mycorrhizae [AM]), are much less conspicuous and have not been studied as extensively as ectomycorrhizae. However, they affect many more plant families and are probably more ecologically significant. These fungi invade plant roots and dwell within cells, much in the same fashion as many obligate fungal parasites (plate 145).

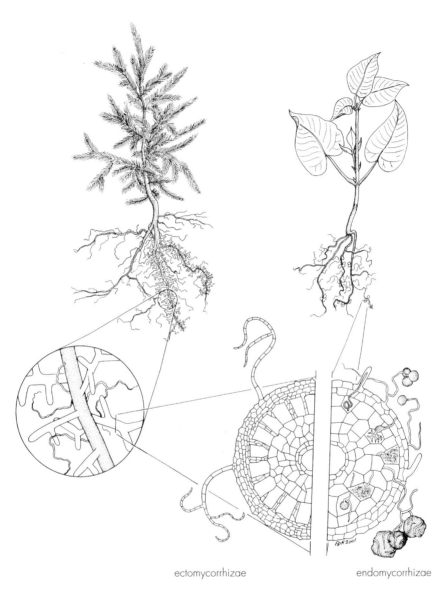

ectomycorrhizae   endomycorrhizae

Figure 10. Ectomycorrhizae (left) and endomycorrhizae (right), shown with representative plant hosts—a conifer and a bean plant, respectively—form distinctively different structures when they colonize roots. Endomycorrhizal (VA) hyphae form a variety of often elaborate structures known as vesicles and arbuscles, shown inside host plants cells; these structures facilitate the exchange of water and nutrients between fungus and plant. Both ecto- and endomycorrhizae reproduce by means of large, globular spores attached to their hyphae in the soil.

The only external sign of infection may be near-microscopic fungal filaments that exit plant roots and run out into the soil, where they absorb water and mineral nutrients for transport back to the roots. Within the plant roots, endomycorrhizae digest their way through plant cell walls and send specialized hyphal structures into plant cells. These structures (known as vesicles and arbuscules) resemble the haustoria formed by obligate plant parasitic fungi (see the preceding section on parasitism); vesicles and arbuscules differ mainly in that there seems to be a reciprocal exchange of nutrients, not a unilateral theft as with haustoria.

Plant hosts seem to react to invasion by endomycorrhizal fungi much as they do to invasion by fungal parasites, and plants seem to tolerate the fungi only as long as they benefit from them. Under conditions of high phosphorus and water availability (as in many heavily fertilized and watered gardens or agricultural fields), the mutualism can break down and the plants may evict the fungi and become cured. If plants are unable to photosynthesize enough carbohydrates to supply their own needs as well as those of their erstwhile fungal partners, however, the mycorrhizal fungi may become parasitic and stunt the plants' growth as they steal carbohydrates. Growing mixes inoculated with spores of endomycorrhizal fungi are now commercially available, but it remains to be seen whether plants will significantly benefit from artificial inoculation of these fungi in greenhouses or gardens. For many years foresters and arboriculturists have been artificially inoculating tree seedlings with ectomycorrhizae to enhance survival and growth as the seedlings are transplanted into the natural environment from more protected nursery beds.

Lichens
Lichens are associations of fungi and algae commonly found on rocks, bare soils, tree branches, and other exposed, nutrient-poor locations (plate 146). They represent one of the most complex and highly developed mutualisms. Lichens are so different in form and function from their component partners as to essentially constitute distinct organisms, and they are classified into unique orders, families, genera, and species. In fact, the component organisms—particular species of fungi, green algae, and blue-green algae (cyanobacteria)—are usually not found free-living, and most appear incapable of reproduction or even survival in the absence of their symbiotic partners. In lichens, the fungal filaments and algal cells become entangled and, in many cases, establish direct contact between their cell mem-

branes (plate 147). This intimate contact greatly facilitates reciprocal exchange of nutrients, hormones, and molecules jointly synthesized by the two organisms.

What are the benefits to the partners in the lichen symbiosis? The algae appear to receive physical protection from the environment by the tough, dense, pigmented fungal tissues. The algae also gain access to moisture and mineral nutrients, which lichen fungi are extraordinarily capable of extracting from dry or frozen soil, rocks, or even the atmosphere. In turn, the fungi gain carbohydrates from the photosynthetic algae, as well as available nitrogen from blue-green algae. Both partners cooperate in synthesizing a wide range of organic compounds known to exist in no other living organisms. These lichen substances seem to protect the exposed but very slow-growing and long-lived lichens from herbivores and parasites and possibly function in obtaining mineral nutrients and water from nutrient-poor environments.

Lichens can survive in harsh environments that sustain no other macroscopic life forms, and they may live for thousands of years. Some lichens on pieces of wood incorporated into Medieval or Renaissance buildings or enclosed in glass cases have even continued growing for hundreds of years with no apparent source of moisture or nutrients aside from the indoor atmosphere!

Exactly how lichens reproduce is a mystery. Lichenized fungi freely release viable spores, and lichenized algae can divide and shed free-living cells into the environment, where they may independently exist for a time. However, no one has ever seen the spontaneous formation of a lichen from its separate partners, nor fully succeeded in artificially forming a stable lichen, despite many attempts. Nevertheless, lichens do reproduce and disperse themselves in nature—sometimes over great distances.

Like all mutualisms, the lichen symbiosis appears to be a mutually beneficial exploitation and the relationship can break down under conditions that selectively favor or harm one partner or the other. When low light intensity impairs photosynthesis by the algae or when the fungus is able to obtain abundant carbohydrates from elsewhere, the fungus may injure or kill its associated algal cells. Conversely, if moisture and soluble mineral nutrients are readily available in the environment, some algae may divorce their fungal partners and take up a free-living existence, at least briefly. Such conditions are rare in nature, however, and lichens rate as one of the most widespread—certainly one of the most conspicuous and colorful—examples of mutualism.

Symbiotic Rhizosphere Microbes

Large numbers of fungi, bacteria, and actinomycetes that obtain nutrients from plant exudates or sloughed-off tissues dwell in the soil immediately surrounding plant roots—the rhizosphere. Some of these microbes appear to benefit plants by producing antibiotics, plant growth hormones, or other substances that protect plants from pathogenic microbes. The best-known and most widely commercialized of these organisms is the bacterium *Agrobacterium radiobacter* K84, which produces substances that inhibit its pathogenic relative *Agrobacterium tumefaciens* and thus prevents crown gall disease of plants such as peaches and grapes. Other beneficial rhizosphere microbes, mostly bacteria in the genera *Pseudomonas* and *Bacillus* and fungi of the genera *Trichoderma* and *Trichophyton*, have been identified and attempts made to commercialize them as protective seed treatments or soil inoculants. Thus far, however, these artificial inoculants have achieved only variable success under field conditions.

Endophytic Fungi

A plant-microbial mutualism of mixed consequences for human beings is the partnership of fescues and a few other genera of cool-season grasses with the endophytic fungus *Epichloë typhina* (known as *Acremonium coenophalium* in another phase of its life cycle) and, possibly, other fungi (plate 148). The fungus invades a variety of host plant tissues, including seeds, much like an aggressive parasite. However, endophyte-infected plants appear to be significantly more resistant than are uninfected plants to drought, nutrient deficiencies, and herbivory by nematodes, insects, and mammals. Under some circumstances, it is this resistance to mammalian herbivory that creates problems for humans. Sheep, cattle, and horses that feed on endophyte-infected grasses may lose weight, become nauseated, fail to conceive, abort unborn fetuses, become lame, develop eye problems, or come down with a spastic paralysis. From the perspective of the plants, reduction of herbivory is certainly beneficial. However, sheepherders, cattle ranchers, and horse fanciers are not so pleased, and extensive efforts have been devoted to producing endophyte-free fescues for agricultural use. In contrast, fine-leaved strains of fescues have been deliberately inoculated with endophytic fungi and marketed as pest-resistant and drought-tolerant turf-grasses for lawns, golf courses, and public parks.

Plant-Animal Mutualism

Insects and birds are sometimes highly selective in their choice of the species that they pollinate. Many plant species, in turn, are most effectively pollinated by particular species or genera of pollinators. Because plants can be highly dependent on insect pollination to reproduce and complete their life cycles and pollinating insects require pollen and nectar for their own survival, such a relationship is indeed a mutualism, even though the participants may dwell far apart.

An extreme example of a plant-insect pollination mutualism is that between certain Middle Eastern species of figs and caprifig or agaonid wasps (family Agaonidae or Torymidae). These fig trees produce no flowers in the familiar sense of the word. Pollination can only be accomplished by female wasps that visit the male wasps, which dwell within the male sex organs of the trees. After visiting the male wasps, the females, now bearing both wasp sperm and fig pollen, burrow into the female ovaries of the fig trees, where they lay their fertilized eggs and pollinate the plants at the same time. In other cases, the male wasps visit the females. The plants cannot set fruit without the wasps, and the wasps cannot produce young without both male and female or hermaphroditic fig trees—an obligate mutualism if ever there was one. Other insect-plant pollination relationships are not so stringent, but it is well known that, for instance, honey bees have strong preferences for pollinating certain types of flowers, bumble bees others, and syrphid flies yet others.

Another highly specific plant-animal mutualism involves certain aggressive South American and African ants of the genus *Pseudomyrmex* and thorny shrubs (mostly *Acacia*). Although details of the New and Old World symbioses differ somewhat, they share important features. In the South American version of the association, the primarily carnivorous ants excavate tunnels and chambers in the plants' large thorns, which are hard-shelled but hollow or soft-pulped. Inside these armored chambers, the ants reproduce, store insects and other prey that they have killed, and gain shelter from the weather and protection from their own predators (mostly birds). However, should an herbivore—a hungry insect or a cow—disturb the plant by attempting to eat leaves or other plant tissues, the ants swarm out from their thorny fortresses to bite and sting the intruder until they kill it or drive it away. As a result, the shrubs remain remarkably free of herbivore damage in regions swarming with herbivorous insects, birds, and mammals. Acacias artificially cured of their ant partners by means of insecticides are soon defoliated by herbivores despite their thorns—strong evidence of the effective defensive role of the ants.

Although best known in the tropics, ant-plant mutualisms also exist in temperate zones. Black cherry supports ants with extrafloral nectaries, and in turn receives some protection against herbivores such as the eastern tent caterpillar (*Malacosoma americanum*).

Animal-Animal Mutualism

There are countless instances in nature of animal-animal mutualisms, but many striking examples such as rhinoceros (family Rhinocerotidae) with tickbirds (*Buphagus*), crocodiles (family Crocodylidae) with sentry birds, and baleen whales (suborder Mysticeti) with baleen-cleaning fish are far removed from the average garden. However, one animal-animal mutualism that *can* play a significant garden role is that between certain ants and aphids (plate 149). Aphids can be serious garden pests, as they debilitate and stunt plants and transmit viral diseases. Aphids are also slow-moving, soft-bodied, nutrient-rich creatures preyed on by numerous predators and parasites ranging from birds to parasitoid wasps. To the benefit of the rather defenseless aphids, a number of aggressive biting and stinging ant species often adopt them and protect them from their natural enemies, while herding them between leaves and even from plant to fresh plant, much like a rancher moving livestock from pasture to pasture. What the ants gain from the association is sugar-rich honeydew, a secretion of the aphids that the ants actively "milk" from the posteriors of their livestock.

Microbe-Animal Mutualism

Without the aid of symbiotic gut-dwelling microbes, many herbivorous animals would probably starve—no matter how much plant material they consumed. Of course, this would ultimately greatly reduce herbivory of plants. Microbes dwelling in specialized digestive organs of ruminant mammals such as cattle, goats, sheep, and deer enable these herbivores to derive nutrition from many plant materials that otherwise they could not digest (plate 150). The microbes accomplish this feat by digesting plant materials such as cellulose and hemicelluloses that mammals cannot break down. Some mammals that are not ruminant (horses and rabbits, for instance) similarly benefit from the activities of gut-dwelling microbes. These microbes also synthesize essential vitamins and amino acids that are absorbed by their host animals. Benefits to the microbes include an abundant supply of food; a moist, warm, sheltered, stable environment; and the near-absence of predators (other than the host animal itself, which may continually digest large numbers of the gut microflora). In recent years, livestock farmers and horse owners have become aware of the importance of avoiding antibiotics that upset the gut microflora of their animals (much as antibiotic treatments may cause diarrhea in humans). Many animal raisers now actually feed their animals probiotics, commercially produced cultures of live gut microflora or substances that enhance growth of beneficial gut microbes.

It is less well known that somewhat similar microbial mutualisms aid many insects, the most voracious herbivores of all, in digesting their meals. One especially fascinating mutualism is the three-level association among wood-eating termites, protozoa in the termites' guts, and bacteria inside the protozoa. Termites cured of their microbial partners by means of antibiotics starve to death, no matter how much wood they ingest. Another potentially significant mutualism, this one between certain nematodes and bacteria, has recently been recognized. Some nematodes of the genera *Heterorhabditis* and *Steinernema* bore into soil-dwelling insects such as grubs of squash vine borers (*Melittia cucurbitae*) and Japanese beetles. Much like torpedoes, these nematodes pack warheads of insect-pathogenic bacteria that they release into the guts of the penetrated insects. The bacteria immobilize or kill the insect larvae; the nematodes then grow and reproduce, feeding off the insect tissues, and finally pick up a fresh charge of bacteria before they go off in search of new victims. Such loaded nematodes are now available from commercial firms that sell biological control agents. Both bacteria and nematodes clearly benefit from this symbiotic association—although the insects certainly don't.

⊷⊶

In the final paragraph of his most famous book, *On the Origin of Species*, Charles Darwin described the natural world as "an entangled bank, clothed with many plants of many kinds, with birds singing on the bushes, with various insects flitting about, and with worms crawling through the damp earth." Gardens, although perhaps simpler than most natural ecosystems, are entangled banks nonetheless. Plants may be the most prominent inhabitants of gardens, but they are surrounded by a multitude of animals, fungi, and microorganisms. All of these organisms interact in countless ways, both beneficial and harmful, and the ecologist-gardener who understands these actions can use this knowledge to practical advantage and is likely to have a richer and more interesting gardening experience. In the next chapter, we describe how gardeners can actively promote or discourage particular garden organisms or interactions between organisms for the benefit of the garden ecosystem.

6

Gardening as Applied Ecology

In many ways, the term *natural gardening* is an oxymoron. By "natural" we tend to mean unaltered by humans, but we humans are a critical and inseparable part of garden ecology. Gardening is most often an attempt by humans to get plants to grow in places or in ways that they would not if left to their natural tendencies. Indeed, many modern garden plants would not survive for long if left untended by humans. Many others would revert in a few generations to ancestral forms often quite different from those found in modern gardens. Nevertheless, skillful application of ecological principles can enable the wise gardener to work with Nature, rather than constantly engaging in an expensive—and often losing—battle against it.

Responsible gardeners should regard themselves as stewards of the Earth and its inhabitants, as persons entrusted with their care and entitled to reap reasonable compensation for services provided—yet accountable for their management to neighbors, future generations, and other living organisms. First, gardening practices should do no harm, for example, by not aggravating soil erosion, not depleting soil organic matter in excess of its rate of formation, or not applying pesticides in amounts that exceed their rate of degradation to harmless materials or in ways that harm beneficial or harmless organisms. Gardeners should also adopt positive practices such as enhancing soil organic matter formation and pesticide degradation, attracting and sustaining beneficial organisms, and selecting and managing plants to minimize resource consumption.

A comprehensive discussion of techniques for sustainable gardening or wise stewardship of the land would require an entire library of how-to gardening texts and is beyond the scope of this book. However, the following examples of garden management strategies and techniques illustrate how a conscientious and responsible gardener can practice good stewardship of the land and wise management of the cast of characters in the garden drama. These suggestions may also provoke further thought, investigation, and experimentation.

Stewardship of the Soil

Generating and maintaining a high level of soil organic matter is one of the most effective tactics for ensuring a successful gardening experience. Abundant soil organic matter loosens and aerates tight clay soils, tightens up excessively sandy soils, absorbs and retains large amounts of water and plant nutrients, acts as a slow-release reservoir of nutrients, adsorbs and neutralizes pesticides and other toxic substances, promotes growth of desirable microorganisms, buffers soil pH, helps prevent soil compaction, and generally benefits virtually every other aspect of the garden environment.

Another key element in wise soil management is controlling erosion, because erosion steals the richest part of the soil and turns it into an environmental pollutant—a double sin. Reducing surface erosion of soil may be accomplished by planting strongly rooted, soil-retaining perennial plants on steep or long slopes or by avoiding gardening on them altogether (plate 151). Soil erosion can also be reduced by building terraces across slopes and by planting strips of dense perennial vegetation between tilled garden spots. If possible, builders of new homes should have the retention of topsoil specified in construction plans. Unfortunately, due to prior soil erosion or disturbance, many gardeners discover that the subsoil is their only available soil. The remedy? Well, a gardener may face a tedious, expensive job of rebuilding the topsoil with lime, fertilizer, and lots of compost and other organic materials or may need to purchase topsoil taken from somewhere else.

Compacted soils, which may result, for instance, from foot or vehicle traffic or application of salts, are hostile to most plants. Only specially adapted species can thrive in such soils. Common and buckthorn plantains (*Plantago major* and *P. lanceolata*, respectively), widespread weeds that tolerate anaerobic soil, were once supposedly known by Native Americans as "white men's footprints" because they tended to grow where heavy, high-heeled riding boots crushed the soil. They may still serve as useful indicator plants of soil compaction in gardens or lawns. Indicator plants for soil compaction may vary by locality, but other widespread North American indicator plants include heal-all (*Prunella vulgaris*) and nutsedge (*Cyperus esculentus*). Some European examples include hawkbit (*Leontodon*), docks (*Rumex*), and dandelions.

Farmers and gardeners sometimes unknowingly create hardpans by driving vehicles across soil, plowing or rotary tilling at the same depth repeatedly, or by walking on damp ground (a footstep can exert as much or more force per unit area as does a tractor's tires). Generally, more than two-thirds of all eventual soil compaction damage occurs on the first impact. The effects of a single pass by an all-terrain vehicle over a hillside on Steve Salt's farm in 1994 were still apparent eight years later. Thus, if you *must* walk in your garden, establish permanent walkway zones or lay boardwalks and stick to them. Walking in a random fashion in a garden in the mistaken belief that little damage will be done in any one spot will just compact the soil everywhere.

Other measures to reduce compaction include staying off soil when it is wet and building up organic matter to make the soil more spongy and resistant to compression. A plowpan (artificial hardpan) can be very difficult to remedy, requiring strenuous, expensive, and time-consuming treatments such as deep digging or forking, deep chisel-plowing, or, in severe cases, subsoiling with a monstrous spike dragged through the ground by a bulldozer. If soil compaction is aggravated or caused by sodium salts, physical remedies alone are inadequate. Large amounts of calcium and/or magnesium (such as gypsum or Epsom salts) will have to be worked into the soil, and remediation of the puddled soil will be a slow, difficult process of doubtful prognosis. Prevention is the best remedy for soil compaction. Take care of your soil and it will take care of your plants—which will then take care of you.

Stewardship of the Water

There is a great deal of water in the world, more than enough to meet projected needs for the indefinite future—if it were all useable. However, water sufficiently free of harmful contaminants to be suitable for drinking or irrigation is a scarce and dwindling commodity. Many environmental scientists hold that it is shortages of high-quality water, not of food or energy, that may ultimately limit the growth and well-being of the human population. Agriculture is the greatest consumer of water among human activities, and home gardening and landscaping frequently consume several times as much water per area of land as does commercial agriculture. The price of irrigation and drinking water in most areas has never reflected its true costs, but the era of cheap water may be vanishing. Thus, responsible gardeners should consider ways to increase efficiency of water usage in their gardens (plate 152).

Not much can be done by an individual gardener about the quantity or quality of rain or snow that falls on a garden. Gardens virtually everywhere suffer from periodic dry spells and in some regions only desert plants can be raised without regular watering. Leaving aside a debate about the ethics of using irrigation to raise exotic plants in the desert, there are tactics and methods that can be adopted to minimize dependence on irrigation and enhance its efficiency (plate 153). Some of these techniques include choosing species or varieties of plants that are drought-tolerant (xeriscaping); grouping plants with similar water needs and tailoring irrigation to the differing needs of garden zones; applying mulches to reduce evaporation of water from the soil; shading plants and creating windbreaks to reduce transpiration; watering at cooler times of the day to minimize wasteful evaporation; using water-conserving technology such as trickle or drip irrigation; and adding organic matter to soil to increase its water-holding capacity. In some circumstances, excessive water can be a problem in a garden—one that may be more difficult to deal with than insufficient water. Possible solutions include building raised beds or digging drainage ditches (plate 154), or if the situation is chronic, planting wetland-adapted plants.

Other good water management practices include maintaining soil moisture during winter for dormant perennials, whose roots may not be dormant at all, by continuing irrigation, if necessary, in dry autumns after leaf fall until the ground freezes; avoiding contamination of groundwater with fertilizers or pesticides; avoiding use of salty water; recycling gray water, household waste water that may contain biodegradable soaps and common dirt, but that is free of excrement or harmful chemicals; and capturing runoff from roofs and paved areas in cisterns.

Stewardship of the Atmosphere

Responsible and intelligent gardening and farming can actually help improve air quality. Healthy vegetation mops up greenhouse gases and other atmospheric pollutants, generates a net output of oxygen, and buffers against extreme swings in air temperature and humidity. Large deciduous trees strategically placed to shade yards or houses can reduce daytime air temperatures as much as 8–10°F (5–6°C) and, combined with appropriate shrubs, can reduce summer cooling costs by up to 50 percent. One commonly cited figure is that one large deciduous tree may be worth about 5 tons (60,000 Btu/hour; 17.5 kW) of air-conditioning capacity. In winter, conifer windbreaks combined with evergreen foundation shrubbery or espaliered vines can reduce wind velocities up to 50 percent and heating costs by 10 to 25 percent. Trees and shrubs may also strongly reduce noise, odor, and dust pollution. Species or varieties of plants suitable for pollution and temperature control vary greatly from one location to another, however, and it would be wise to consult a specialist or at least to obtain literature on the subject from your local extension service or equivalent agency.

On the do-no-harm side of the ledger, garden power equipment should be used no more than necessary and kept well maintained to minimize noise and noxious emissions. Replacing power equipment with good-quality hand tools would not only be feasible in many smaller modern yards, but provide a valuable antidote to the health damages of sedentary lifestyles—as well as save money and reduce environmental damage. Work out with a push mower and re-acquaint yourself with the unique sound of a reel! Alternatively, replace high-maintenance turfgrass lawns with locally adapted low-maintenance ground covers, and retire your lawnmower altogether. If you deem it necessary to apply pesticides or other substances, minimize applications and avoid spraying when it is windy or hot to help reduce air pollution and contamination of nontarget areas or organisms. Also, minimizing generation of dust (such as by not tilling overdry soil) will not only improve air quality, but also reduce soil erosion.

Managing Garden Organisms

Gardening is concerned with enhancing the growth of desirable plants and doing battle with their enemies. Often it seems that the bulk of gardening

advice deals with pests and how to reduce or eliminate them. From an ecological perspective, however, it may not be desirable, even if feasible, to
utterly eliminate garden pests. Instead we should try to manage both pests
and their often more numerous desirable garden neighbors for the greater
welfare of the garden and the environment as a whole.

Plants

The choice of which plants to raise rivals care for the soil, water, and air in
its effect on the environmental impact of gardening, as well as on its success. For example, attempting to raise water-loving plants in dry climates
or upland plants on soggy wetland sites necessitates an expensive and frequently ecologically disruptive struggle. So does attempting to grow heat-
loving plants in cold climates or disease-susceptible plants in regions with
high levels of pathogens. Within particular plant species, some varieties
may be less susceptible than others to disease, thriftier with water or nutrients, more tolerant of cold or heat, and so forth. Carefully selecting well-
adapted plant species and varieties is likely to greatly reduce the resource
consumption, expense, and work involved in gardening, as well as minimize undesired environmental damage. But when all is said and done, even
environmentally conscious gardeners may want to grow more than cacti
and mesquite (*Prosopis*) in Arizona, temperate rainforest plants in coastal
Washington, or warm-season grasses in Nebraska, so environmental re-
engineering is likely to remain a major gardening activity. It's well to
remember, though, the less remodeling of the garden site required, the less
the expense and ecological disruption. And choice of plants will be a primary determinant of how much remodeling will be required.

Some plants enter the garden uninvited. Unwelcome plants (weeds)
compete with desired plants for nutrients, space, and light; may carry disease organisms or pests; or sometimes actually poison nearby plants.
Controlling weeds occupies much of the time, effort, and expense invested
by many gardeners. However, for environmentally responsible gardening, it
is crucial to keep in mind that the goal of weed management is to minimize damage to desirable plants, *not* to destroy plants that the gardener
did not intentionally plant. Weeds should be *controlled*, not necessarily
exterminated. Also, some weed-control materials and techniques, if inap-
propriately used, may damage desirable plants, contaminate the environment with toxic chemicals, injure beneficial organisms, destroy soil organic

matter, or compact the soil—all of which may cause more harm than the weeds themselves.

There are ecologically sound ways to prevent the growth of weeds, including the use of mulches or landscape fabrics to smother weeds, flaming young weed seedlings to create stale seedbeds, densely planting desirable plants (keeping in mind, however, that this will increase requirements for water and nutrients), selectively watering and fertilizing only desired plants, and applying corn gluten around established or perennial plants to inhibit germination of annual plant seeds. Methods for controlling established weeds may include careful hand-pulling or tillage, done so as not to injure desirable plants or to cause soil compaction; flaming; or, as a last resort, using the least toxic and most rapidly biodegradable herbicides available—such as acetic acid (concentrated vinegar) or fatty acids—applied so as to minimize environmental contamination and contact with desirable organisms.

Vertebrates

Getting along with birds, dogs, rabbits, deer, moles, and other garden visitors can be a real challenge for gardeners, especially in suburban and rural areas. But sometimes gardeners also wish to encourage wildlife to take up residence. The basic key to encouraging animal visitors to gardens is providing food, shelter (such as brush piles, weedy patches, or abandoned containers or sheds), and water and eliminating barriers or hazards such as predators. For example, toads voraciously consume many harmful insects; they may be encouraged to take up residence by providing wood piles or toad houses and shallow ponds or trays of water. In tropical areas, geckos and lizards fulfill a similar role in gardens and may be likewise encouraged.

Of course, gardeners often invest energy in trying to keep wildlife *out* of the garden. However, traps, poisoned baits, odoriferous repellents, noisemakers, or the like rarely deter unwanted animals for long. (See chapter 2 for some creative—but often largely futile—methods that have been used by gardeners in attempts to deter moles, rabbits, rodents, and deer.) Barriers may reduce visits by animals but are almost never totally effective. Those barriers that do approach that ideal are generally expensive, elaborate, and/or unsightly—and may pose impediments or hazards to human visitors. Because, almost by definition, gardens are food sources for animals, eliminating food may necessitate eliminating the garden or at least severely limiting what plants may be grown. In other words, gardens—especially those located on the borders of wild areas—may receive visits by wildlife no matter what the gardener does. Thus, gardeners may need to make decisions about what level of loss to wildlife is tolerable.

A goal for many gardeners is attracting hummingbirds, purple martins, swallows, flycatchers, and other birds to sing, entertain with their antics, and/or consume harmful insects. Of equal interest is reducing the losses to hungry birds of fruits and newly planted seeds and repelling starlings, brown-headed cowbirds (*Molothrus ater*), and other bullies of the bird world. How can these apparently conflicting goals be reconciled in an ecologically sound fashion? Well, they probably can't—at least not completely—although skillful garden management may enable some degree of compromise. If one's primary goal is attracting birds, then growing plants that have attractive or nutritious seeds or fruits or that provide perching or nesting sites is a good strategy, along with more direct approaches such as installing and maintaining bird baths, bird feeders, or nest boxes designed for particular birds and keeping domestic cats out of the garden. Many nineteenth-century American farmers grew, dried, carved, and hung birdhouse gourds (club- or dumbbell-shaped varieties of *Lagenaria siceraria*) to provide nesting places for martins or wrens (family Troglodytidae). Bird lovers may take encouragement from the observation that in temperate climates, the highest bird populations and greatest species diversity are frequently found not in wild areas, but in landscaped suburban areas.

Food is a major incentive for wildlife to visit gardens, and some gardeners raise plants especially for that purpose (plate 155). State or local conservation or natural resource agents and local birding and gardening groups can provide advice as to the best species to plant in different regions. Low-maintenance plants in the Lower Midwest that attract birds (and often other wildlife as well) include choke cherries (*Prunus virginiana*), mulberries, serviceberries, barberries, butterfly bushes, eastern red cedar, *Viburnum* species, spicebush (*Lindera benzoin*), chicories, and various large-seeded grasses. Unfortunately, sweet corn, sunflowers, rose family fruit trees, and cole crops such as cabbage and broccoli are also attractive to a wide range of wildlife.

Some species of birds are normally quite destructive to garden plants, whereas many birds are only destructive under certain circumstances. Starlings, sparrows, quelas, and blackbirds are almost always unwelcome during the growing season. Tactics against garden-pillaging birds include enclosing trees and bushes in mesh netting (plate 156); hanging bright-colored, reflective, fluttering, or noisy objects (generally only temporarily effective); playing loud music or recordings of bird distress calls; or planting attractive sacrifice crops such as mulberries to divert birds from more valuable plants such as cherries. And what can one do about plundering of newly planted seeds by such birds as blackbirds and crows? Conventional (nonorganic) gardeners sometimes use fungicide-treated seeds, which, in addition to being protected from soil molds, often taste bad to birds. Other—often impractical or ineffective—defensive methods include laying down mesh over planted areas, standing guard, posting a watchdog, or planting substantial extra seed. However, there is one practice to absolutely avoid, regardless of your gardening philosophy: carelessly planting seeds so as to leave some lying unburied on the soil surface. Unburied seeds will be promptly spotted by birds and will serve as an invitation to search for other buried seeds nearby.

Insects

Insects are probably considered the worst pests in most gardens, with many gardeners viewing *all* insects as mortal enemies. However, a great many insects are highly beneficial (see chapter 2). Before attempting to control insects, a gardener should find out just which insects are present in a garden and what they are doing (plate 157). Sources of information on local insects can include cooperative extension services, university entomologists, or printed or Internet field/garden guides to insects. Keep in mind, though, that insects tend to have definite geographical distributions; also, species common in the wild may be scarce in yards or gardens and vice versa. Thus, a field guide useful for one region or country may be of little value in another, and field guides may have limited application in gardens.

As with other animals, attracting and nurturing beneficial insects depends first on recognizing them and their particular needs and then providing food, water, and shelter from their enemies. Especially desirable in most gardens are European honey bees and other pollinating bees and wasps (plate 158). Unfortunately, parasitic tracheal and varroa mites that arrived from Asia in the 1980s have decimated most North American honey bee hives and forced gardeners to turn to other bees for pollination. Some of these other bee species are effective pollinators, but most are solitary rather than colony-forming, and none produces useable amounts of honey. Many of these other bees live in wild areas near gardens and can be encouraged to move closer and visit garden plants by providing ample nectar sources and specialized shelters such as the orchard mason bee (*Osmia lignaria*) houses now sold through gardening catalogs. Other pollinating insects include flies (even some otherwise obnoxious housefly species), beetles, butterflies, and moths.

Beneficial insects also include many predators and parasitoids (discussed in chapters 2 and 5 and in more detail later in this section). These insects help control a wide range of organisms that prey on plants. But how can a gardener encourage such organisms without also promoting populations of plant-devourers? One answer is that—if present in nearby weedy or wild areas and if not eradicated by nonselective pesticides—these beneficial insects will almost automatically immigrate and increase as populations of their prey increase. Also, adult forms of many parasitoids are actually nectar-drinkers, and they can be attracted and sustained by planting food-source plants. For example, adult tachinid flies feed on floral nectars of celery family plants like Queen Anne's lace, dill (plate 159), and ammi majus, whereas brambles provide nectar for *Anagrus epos* wasps, whose larvae parasitize eggs of leafhoppers.

How can gardeners control damaging insects without injuring themselves, beneficial organisms, or the environment? As with weeds, the emphasis is on *control*, not elimination. Attempts at eliminating insect pests, besides usually being futile, may do more harm than good. Indiscriminate spraying of nonselective pesticides often kills as many or more predators of harmful insects than it does harmful insects. Second, it's important to decide if control is actually needed at all; an occasional bite out of a leaf or fruit really does very little harm to either the plant or the gardener, and waging warfare on the supposed culprits may be a waste of time and money—or worse. Third, if truly significant and intolerable harm is being done to garden plants by herbivorous insects, it's important to accurately identify the guilty parties and take steps to combat them selectively, not to massacre innocent bystanders. The garden environment can be managed to enhance populations and activities of beneficial parasitic and predaceous organisms.

Unfortunately, use of insecticides is the first—and often the only—tactic many gardeners employ, when it really should be the last.

Insecticides are often applied at unwise times (such as during winds or rain), in excessive amounts, or in such a way that most of the applied material ends up contaminating nontarget areas. Also, repeated applications of the same insecticide may select for resistance in insects, so that the chemical gradually loses its efficacy.

Nevertheless, there may be occasions when insecticides are called for. Sudden or overwhelming insect outbreaks that cause severe plant injury may yield to nothing else. In such cases, the gardener should carefully identify the culprit insect(s), select an appropriate insecticide in proper dosage, use protective equipment (gloves, apron, goggles, and respirator mask, as appropriate), and apply the insecticide in such a manner as to effectively treat the target species while minimizing injury to nontarget organisms. Insecticides should be applied to both top and bottom surfaces of leaves (many pests hide underneath). They should not be applied right before a rainstorm, which will simply wash them off. Spraying or dusting should not be done when it is windy nor, if possible, in the morning (when pollinating insects are often most active). Bright sunlight or high temperatures sometimes cause insecticides to harm plants or create fumes that may drift and affect nearby unsprayed plants—and people. Some pesticides such as summer oil, insecticidal soap, sabadilla, neem, rotenone, pyrethrins, and *Bacillus thuringiensis* (Bt) are considered organic or natural. However, "natural" does not necessarily equal safe, either to a gardener or to the environment. Natural materials should be handled with the same care as synthetic chemicals. Most of the information necessary for safe and effective use of an insecticide is found on its legally required label. Read the directions!

The identification and synthesis of insect developmental hormones and hormone-antagonists (substances that interfere with hormonal activities) are recent developments that may lead to safer, more selective chemicals for controlling pests without harming nontarget organisms. Treating insects with these substances rarely kills them outright; instead they do not pass through metamorphosis properly and thus fail to reproduce. Also, because insect hormones tend to be highly species specific, using them may have little impact on nontarget organisms. Another advantage is that insects are unlikely to develop genetic resistance to their own hormones, because this would be suicidal. Development of this potential insect-control technology has been quite limited so far, however, partially due to the high cost and low stability of most insect hormones—but more so, ironically, to the extreme specificity of these chemicals. Few chemical manufacturers are willing to incur the high costs (up to $100 million per pesticide) of developing, testing, registering, producing, and marketing a material that will only affect a very narrow group of insects (or even a single species) and thus have a very limited sales potential.

Other insect control techniques may not be as quick and easy as spraying nor give the spectacular body count of insecticide applications, but if judiciously used these can greatly reduce the need for insecticides. And if used in combination, ecologically oriented measures can often give control (at least at the garden scale) equal to that of insecticides, but without the health and environmental hazards—although they might also require more skill and effort. Gardeners can raise trap or companion crops, modify the timing of planting, create fallow or plant-free periods, maintain good garden sanitation by removing sickly or dead plants and debris, remove pests by hand, blast insects with jets of water, set out lure-traps or colored sticky balls and sticky cards, choose insect-resistant plants, apply nontoxic chemical controls such as repellents, and position physical barriers such as aluminum foil collars and gauzy cloth covers.

A trap crop is a plant species that herbivorous insects prefer over other plants in a garden; it draws the herbivores away from the more desired plants. Insects concentrated on the trap crop can then be destroyed by mechanical means or by localized pesticide applications. For example, zucchini (*Cucurbita pepo*) tends to draw cucumber beetles and squash bugs away from many other cucurbits; alfalfa attracts many cotton pests; kochia (*Kochia scoparia*) is strongly attractive to grasshoppers, as is spearmint. Similarly, radishes will draw wireworms (larval stage of click beetles) away from carrots, and smartweed (*Polygonum* or its domesticated relative water pepper *Persicaria hydropiper*) will lure Japanese beetles away from most other garden plants.

When planted in close proximity to another species, certain plants (companion plants) either repel harmful organisms or attract beneficial organisms. The strong scents of garden plants such as brassicas, umbellifers, and alliums seem to mask the more delicate aromas of other species nearby, thus effectively hiding them from scent-dependent insect pests. Common examples of companion planting include growing sweet basil (*Ocimum basilicum*) with tomatoes or other plants; tansy (*Tanacetum*) or catnip (*Nepeta cataria*) with zucchini; radishes with cucumbers; and certain species of marigolds, such as *Tagetes patula* (French dwarf marigolds), with virtually any other plant. However, the efficacy of companion planting is erratic, varying widely with conditions, location, and particular plants involved, so gardeners must carefully evaluate the success of this strategy in their own gardens with their own plants. Many insect and other pests are specialist herbivores, organisms that selectively feed on one or only a few species or varieties of plants. Thus, the very diversity of plant life in most home gardens will greatly reduce insect predation compared to that seen in monocultures (large areas planted to just one crop), such as farm fields, lawns, and golf courses. Insects may have a more difficult time finding their preferred food source among a tangle of diverse species and can rarely multiply to large populations with the resulting limited food resources. However, if even a few pioneer individuals of well-adapted herbivores locate a large, uniform food source, they may rapidly multiply into a hard-to-control population and do great damage.

Gardeners should be aware of when overwintering insects emerge from dormancy to lay eggs or when young insects disperse in search of new hosts. The planting of host species may be delayed until after these times, when the pest has either starved or emigrated in search of food. For example, a gardener who knew that periodical cicadas (*Magicicada*) were to emerge in the Midwest in the spring of 1997 could have avoided damage to young perennial plants by delaying planting of woody perennials until autumn. If autumn frost dates permit, delaying planting of pumpkins until well after the emergence of adult squash bugs from overwintering hibernation will greatly reduce losses to these insects because they will have already mated and migrated elsewhere in search of food and egg-laying sites before plants emerge. Alternatively, plants can be planted earlier so that they are in a more mature—and usually tougher—condition when insects attack.

Leaving the garden bare (fallowing) can reduce local insect pest populations by starving specialist herbivores or by encouraging both specialists and generalists to seek their dinner elsewhere. A technique related to fallowing, crop rotation (growing unrelated plants in succession on the same piece of ground), often reduces herbivorous insect populations by denying the succeeding generations preferred food sources that nourished their parents. However, this technique is not as effective in small home gardens as in large agricultural fields, because alternative food sources may grow nearby, where insects can conveniently find them by sight or smell.

Garden sanitation—removing dead or diseased plants, litter, and trash and clearing off and destroying all annual plants at the end of the growing season—is very useful in reducing habitat for pests during the growing season and eliminating places to hibernate during the winter. Many insect pests complete part of their life cycle, such as pupation, in stubble or litter during the winter and are then ready to attack young plants the next spring. Burning is the most effective way to destroy insects or insect eggs in trash. If burning is not feasible, high-temperature composting is next best; deep burial is third. Piling plant debris in a corner of the yard or garden is not very effective for eliminating pests, although it is better than leaving it in the garden.

One might question whether keeping a garden "spic and span" might also deny habitat to beneficial insects and other organisms. To a certain extent, this is true, but in actual practice good sanitation usually yields strongly positive net benefits, possibly because herbivorous pest insects generally live in or on host plants, whereas relatively mobile beneficial predaceous or pollinating ones do not, instead residing or overwintering elsewhere. However, providing appropriate nearby habitat (wildflowers, beehives, nest boxes) for beneficials would certainly help them.

Hand-picking can be an effective and highly selective method of insect control, provided one has the time and energy, the garden is small enough, and the infestation is not overwhelming. The insects must also be large or brightly colored enough to be readily seen, be active or visible in the daytime or by flashlight, and move slowly enough to be caught. Some insects can be more easily caught on chilly mornings, when they are more sluggish than at warmer times. Some soft-bodied, relatively immobile insects can be partially or completely controlled by properly adjusted high-pressure jets of water. Mealybugs and scale insects, in particular, are incapable of returning to their host plants under their own power once removed, and aphids are easily crushed.

Lure-traps take advantage of insects' love of certain shapes, colors, or scents. A trap, for instance, may be a container into which they can enter but not leave or an object coated with a sticky goo from which they cannot escape once mired. The attraction of insects to particular colors (see chapter 2) is the basis of sticky traps—colored balls or squares of material coated with a sticky goo that does not dry and is often scented with pheromones (plate 160). Some commercial traps, such as for codling moths and gypsy moths, play an especially dirty trick on the target insects by luring them with come-hither sex attractant pheromones (plate 161). Such traps can be quite selective and may be effective in controlling low to moderate infestations. The main drawbacks are their sometimes high initial costs, their need for maintenance and replacement, and the fact that effective attractants are available for relatively few insect pests.

Breeding plants for resistance to insect predation is not nearly as advanced nor as widespread as breeding for disease resistance, but some insect-resistant species and varieties are available. Farmers in the United States, for example, quite commonly plant varieties of wheat resistant to the Hessian fly. Genetic engineering has been used to develop corn/maize and other crops containing genes of the bacterium *Bacillus thuringiensis* (Bt), a natural pesticide and one of the few acceptable to organic farmers. Some consumers, though, especially in Europe, have expressed fears over environmental and culinary safety of these and other genetically modified organisms (GMOs). (In genetic engineering, an organism's genetic makeup is altered by splicing genes into it from other organisms that may be in a different genus, family, or even kingdom—thus bypassing natural controls on gene flow between unrelated organisms.)

Gardeners also have available to them a variety of chemicals that control insects by means other than poisoning. For example, insecticidal soap, a pure liquid soap without perfumes, colorants, or other additives, kills insects by clogging their breathing pores and smothering them. It can be sprayed on many plants to control soft-bodied insects such as aphids and true bugs. It is also effective against herbivorous mites (which cannot be controlled by most conventional insecticides), but is much less likely to harm hard-bodied insects such as most beneficial insects. Insecticidal soap is nontoxic to mammals and most other animals. Summer oil and dormant oil (also known as winter oil; used only on dormant plants) are highly purified vegetable and mineral oils that are sprayed on plants (usually woody perennials) to smother insect larvae and eggs. Oils are particularly effective against scale insects, which are difficult to control by other means. Both insecticidal soap and oils must be used cautiously, because they injure many plants under cold or hot and sunny conditions and they may harm plants such as umbellifers (carrots, dill, and relatives), conifers, and maples under any conditions. If in doubt, test a soap or oil spray on a small part of a plant and observe the treated plant for several days before using the treatment on a whole planting.

Another nontoxic substance that physically repels or injures target insects is diatomaceous earth (DE). This is a powder mined from ancient deposits of shells of diatoms (primarily tiny marine creatures, which are still common today), and it represents the natural equivalent of ground glass. The powder is dusted on or around plants to be protected, and extremely sharp particles in the powder slice and rip the exoskeleton or mouthparts of insects, causing them to stop feeding and/or bleed body fluids. Diatomaceous earth is effective against chewing or soft-bodied crawling insects. It is even more effective against slugs and snails, lacerating their gliding feet. However, it is important to use insecticide-grade diatomaceous earth (obtained from gardening suppliers), not the grade used in swimming pool filters. Also, it is crucial not to breathe the dust, because diatomaceous earth can tear up lung membranes just as it does insects' shells and skins.

Other nontoxic chemicals may act as repellents by creating offensive odors or flavors that deter insects from feeding on treated plants. These chemicals include natural substances such as extract of hot peppers or artificial chemicals such as naphthalene (used in mothballs). A rather expensive use of nontoxic chemicals is to broadcast strips of paper or other materials impregnated with insect pheromones such as sexual or feeding attractants. The fog of odor drives sensitive insects into a frenzy of frustration and may prevent them from actually finding food or a mate.

Biological control of insects is gaining popularity. In this method, natural predators, parasites, or pathogens of the targeted pest species are introduced into the garden or encouraged to increase, so as to suppress the pest population. Larvae of fireflies or lightning bugs prey on various small insects and snails. Lacewings are voracious predators of soft-bodied insects. Ladybird beetles (or ladybugs) and closely related mealybug destroyers (*Cryptolaemus*, sometimes called ladybugs by commercial suppliers) in both their larval and adult stages prey on a wide range of soft-bodied insects, including aphids, mealybugs (family Pseudococcidae), scales (family Coccidae), mites, and small caterpillars. Syrphid fly larvae attack subterranean larvae of other insects, especially beetles. Yellow-jackets and other vespid wasps paralyze and haul caterpillars back to their nests. Using praying mantises as biological agents has fallen out of favor; although they do attack a large number of pest insects, they also slaughter beneficial insects—and each other!

Parasites that enter and dwell within their target hosts, consuming them from the inside out, are particularly effective control agents. For example, tiny parasitoid wasps, such as ichneumonids, braconids, and chalcid or jewel wasps (family Chalcidoidae) vigorously attack the eggs and larvae of many insect species, such as hornworms and other caterpillars. Other effective parasites include tachinid flies, which primarily parasitize caterpillars, but which also attack larval and adult beetles and young nymphs of grasshoppers and true bugs. The parasitoid wasp *Encarsia formosa* especially has been used for control of aphids and whiteflies in greenhouses (plate 162). The microscopic parasitic protozoan *Nosema locustae* is widely used to suppress grasshoppers, crickets, locusts, and relatives. This protozoan infects susceptible insects, debilitating but rarely killing them, and can be passed to the next generation of insects via egg-laying. Parasitic nematodes are used to control raspberry cane borers (*Oberea bimaculata*), mosquito larvae, and, in strawberries and potted ornamental plants, black vine weevils (*Otiorhynchus sulcatus*). Commercially available bacteria and fungi used in biological control include *Bacillus popillae* and *Bacillus lentimorbus*, bacteria that are applied to lawns to cause milky spore disease of Japanese beetle larvae and other subterranean grubs. Certain strains of *Bacillus thuringiensis* (Bt) are widely used by organic gardeners and farmers to control caterpillars, Colorado potato beetles, flies, and mosquitoes. *Entomophthora* and *Entomophaga* fungi are used to control muscoid flies and locusts.

Although conceptually attractive, the use of biological controls by gardeners is in its infancy and, unfortunately, tends to be rather ineffective (at least outside of greenhouses) as well as expensive. Predators, parasites, and pathogens are all living organisms, each one with its own requirements for food, moisture, temperature, and habitat. Simply releasing organisms obtained from garden supply companies often produces erratic and unpredictable results. For example, mail-order ladybird beetles, which are often commercially collected from hibernating masses of insects in the Sierra Nevada of California, are notorious for promptly emigrating to parts unknown when released into gardens. More success is often seen by improving local habitat and conditions for indigenous predators, parasites, and pathogens of harmful pests and by selective and restricted use of pesticides.

Another means of insect control is the use of barriers to exclude pests from their host plants. Screened cages are effective but expensive. More generally practical is the use of floating row covers (also used to reduce frost damage), nonwoven, porous polyester or polypropylene fabrics that can be laid over plants and fastened down around the edges to exclude flying and many crawling insects (plate 163). Of course, the covers must be removed for at least part of each day (usually morning is best) for insect-pollinated crops, because they are effective at excluding beneficial pollinating bees and flies, too. Other physical barriers include cardboard collars or metal foil wrappings on plant stems to bar insects and slugs from crawling up from the soil, individual bags tied over ears of corn/maize or developing fruits, and fabric or other materials smeared with sticky goo. Regardless of which methods gardeners select to manage insects, it is strongly advisable to vary techniques, both in time and in space. Insects are so numerous and reproduce so rapidly that they can very quickly adapt to almost any environmental challenge.

Arachnids and Symphylans

Spiders, at least in North America, are nearly always beneficial in gardens and should be encouraged at every opportunity. But how does one encourage spiders? By not ripping down their webs, for one thing. Gardeners should also minimize use of insecticides and miticides and not disturb spider egg-cases—usually whitish, fuzzy balls the size of large peas or small marbles lodged in folds of leaves or other crevices. Gardeners can also provide shelter in or near a garden in the form of heavily vegetated areas or hollow structures such as cinder blocks and generally resist the impulse to harass or kill these very helpful creatures.

The primary natural controls of plant-eating mites are the predatory mites. So significant are predatory mites that when nonselective mite-killing pesticides (miticides) were first developed and applied to orchards, infestations by spider and gall mites worsened. It turns out that predatory mites are more susceptible to many miticides than are plant-parasitizing ones, and elimination of predatory mites by spraying led to population explosions of plant-parasitic species. Modern miticide development has focused on materials that selectively control plant-parasitic mites while sparing predatory species. Some predatory mites (members of the genera *Amblyseius*, *Neoseiulus*, *Phytoseiulus*, and *Typhlodromus*) are available commercially for release in orchards or greenhouses to control spider mites and European red mites (*Panonychus ulmi*). Predatory mites may also be encouraged by environmental management, such as increasing humidity, to enhance their activities.

Many mites are highly resistant to most insecticides, and, unfortunately, most effective miticides are exceptionally toxic to humans. But recent studies have shown oil of cinnamon to be miticidal, and it is now being developed in commercial formulations for use as a relatively safe alternative, especially for houseplants. Other control methods for spider and gall mites and other plant-parasitic mites include frequent misting of plants with water (most plant-parasitic mites fare poorly in dampness); pruning and destroying infested plants or plant parts; and spraying with insecticidal soaps, to which most mites are highly vulnerable.

Controlling symphylans is difficult, and, ironically, many organic gardening techniques that involve additions of compost or organic mulches to gardens may aggravate problems with these seldom-seen creatures. Gardeners in areas with symphylan infestations should avoid placing undecomposed materials such as sawdust, wood chips, mature straw, or incompletely rotted compost around garden plants. Other potential control techniques for symphylans include soil-applied insecticides; fallowing soil for extended periods (especially during hot, dry weather); applying chitin-containing materials such as powdered shrimp or lobster shells that promote growth of soil fungi predaceous on arthropods; or, as a last resort, soil-applied insecticides.

Slugs and Snails

Many gardeners know that slugs and snails can do extensive damage to young seedlings and maturing fruits. However, they may be controlled by spreading sharp-edged materials such as diatomaceous earth or broken eggshells about plants or wrapping stems of plants with metal foil. Other helpful measures include encouraging toads or turning chickens or ducks loose in the garden (although chickens, particularly, may end up eating garden plants as well as the slugs). The traditional remedy of setting out shallow pans of stale beer to lure slugs to an alcoholic doom does indeed reduce local slug populations—but may also attract all the slugs in the neighborhood to *your* garden.

Earthworms

To enhance earthworm populations, gardeners should keep soil well watered but also well drained and well aerated. Avoid activities that cause soil compaction. Add plenty of compost or green manure, and cover the surface of the soil with organic mulch, especially in winter. Minimize rotary tillage, which can maim or kill large numbers of earthworms, especially in autumn. Some organic gardeners may unknowingly harm more earthworms by frequently tilling or walking in their gardens than they save by avoiding the use of pesticides and concentrated synthetic fertilizers. Tillage may seem to improve aeration of soil, but it only does so transiently down to perhaps 6–8 inches (15–20 cm), while actually compacting the deeper soil. Better techniques for improving aeration include maintaining high levels of organic matter, avoiding compaction, adding soil structure-strengthening amendments such as gypsum, maintaining good drainage, including deep-rooted plants in crop rotations, and, of course, encouraging earthworms.

Nematodes

Most gardeners are unaware of the great numbers of nematodes found in garden soils. Few convenient methods have been developed for either detecting or managing these organisms, many of which are important predators or parasites of harmful insects and other invertebrates. If a gardener does become aware of these near-microscopic creatures, however, it is most likely because of the effects of plant-injuring species. In some warm regions with sandy soils (such as the Coastal Plain of the southeastern United States), plant-parasitic nematodes may be the most devastating garden pests of all. Once plant-parasitic nematodes become established, control is quite difficult. Prevention is much easier than cure; therefore, great care should be exercised when introducing soil or plants with soil on their roots into the garden; make sure that plants come from state-inspected nurseries or other safe areas. Because some nematodes are also transmitted in seed galls, deformed seeds received from any source should be discarded.

Methods for dealing with established plant-parasitic nematodes include chemical nematicides, biological controls, and cultural techniques. Commercial nurseries as well as farms raising high-value crops (such as strawberries, tomatoes, and bell peppers) in Florida, California, and other nematode-prone areas, have for decades routinely fumigated their soils with highly toxic vaporizing chemicals such as methyl bromide. However, these substances are often more toxic to many of the natural enemies of plant-parasitic nematodes than they are to the nematodes themselves. As a result, after the fumes dissipate from the soil, surviving plant-parasitic nematodes often multiply explosively, leading within a year or two to populations of injurious nematodes even larger than the original populations—necessitating yet another, often heavier, fumigation . . . and so on. Mounting concerns about the environmental and health effects of these substances have led to a ban on the use of methyl bromide and related chemicals in the United States and some other countries.

Fortunately, parasitic nematodes have aggressive natural enemies, most of them soil inhabitants. Best known are fungi that trap nematodes with sticky spores, adhesive knobs or pads, or, most interesting, touch-sensitive nooses that draw tightly closed almost instantaneously when nematodes crawl through them. Fungal filaments then penetrate these trapped nematodes, kill them, and digest their tissues—a turnabout of the technique used by plant-parasitic nematodes against plants. Other enemies of plant-parasitic nematodes include predatory nematodes that stab them with their needlelike stylets; certain amoebas and other protozoans; at least one virus; some earthworms; and at least one flatworm. Unfortunately, attempts to control nematodes by augmenting populations of natural enemies have so far been only marginally successful.

Cultural practices against nematode infestations include mass planting of nematode-antagonistic plants (especially marigolds); using hardwood bark mulch; planting selected nonhost plants that starve nematodes by stimulating hatching of their eggs in the absence of a food source; and fallowing soil for extended periods, which starves some species of nematodes but not others and requires eradication of all volunteer weeds that could serve as alternate hosts. Another technique is the planting of trap crops that attract nematodes to invade their roots. These plants then can be destroyed or allowed to be winter-killed, taking their nematode invaders with them. Repeated rotary tillage may also reduce nematodes, because they, like earthworms, do not get along well with rototillers (although one must weigh the harm to earthworms against the benefits of reducing nematodes). Nematodes cannot survive temperatures above about 130°F (54°C) and can be killed by heating soil or potting mixes with steam (in greenhouses) or by solarization of garden plots (covering damp ground with large, clear plastic sheets during sunny weather), which seems to selectively reduce parasitic and pathogenic organisms rather than beneficial ones. The most convenient solution for many gardeners may be to plant nematode-resistant species or varieties when available; these are indicated in catalogs and on labels by an "N" (plate 164). Unfortunately, many otherwise desirable plants have no resistance to nematodes, and genetic resistance to nematodes is often quickly lost when new races of nematodes arise via natural selection.

Tiny and often ignored, plant-parasitic nematodes are very tough—tough to detect, tough to tolerate, tough to control. Keep them out of your garden in the first place if at all possible.

Fungi

Despite their prominent role as plant pathogens, most fungi are neutral or beneficial for plants. Mycorrhizal fungi (discussed in chapter 5) are among the most beneficial microbial soil inhabitants. They enhance the nutrition of host plants, confer drought resistance, and protect plants from pathogenic soil fungi. Other soil-dwelling fungi do not live in such intimate association with plants, yet benefit them by doing battle with pathogenic fungi, bacteria, and nematodes. Unfortunately, very little is known about managing the soil environment to support these beneficial fungi, although it is likely that enhancing soil organic matter will encourage desirable fungi as it does other beneficial microbes.

Inocula of the beneficial rhizosphere-dwelling fungi *Trichoderma* and *Trichophyton* have become commercially available in recent years for use in treating seeds. For a long time, foresters and tree farmers have been artificially inoculating tree seedlings with mycorrhizal fungi to enhance their survival and growth as the seedlings are transplanted into the natural environment from more protected nursery beds. Growing mixes inoculated with spores of endomycorrhizal fungi are also now commercially available, but it remains to be seen whether greenhouse or garden plants will significantly benefit from them.

Indiscriminate use of fungicides or other fungus-control agents is likely to do as much harm as good, with beneficial fungi being destroyed as readily as harmful. However, failure to provide some measure of protection to plants from fungal pathogens can result in severe losses, especially in mild, damp weather favoring fungal spore germination. Many gardening practices can reduce losses, including spacing, pruning, and supporting plants to maximize free air flow and exposure to sunlight (plate 165) and pruning and removing infected plant parts. Also helpful are reducing the use of sprinklers—especially in the evening or in humid weather; rotating crops; and practicing good garden sanitation—particularly removing, burying, or burning crop residues at the end of the season, as many plant-parasitic fungi and insects overwinter in aboveground plant debris. Mulching to prevent contact of leaves and stems with rain-splashed soil will reduce fungal infections, as will avoiding the use of transplants raised on pathogen-infested ground, planting fungus-free seeds, and raising genetically resistant species or varieties. It is also critical never to use pathogen-infected plant materials for compost; spores that survive the composting process will give diseases a head start in garden areas where the contaminated compost is spread.

Under conditions where environmental practices are insufficient to adequately control fungal diseases, judicious use of chemical fungicides may help reduce losses. Using fungicide-treated seed can greatly reduce losses to damping-off fungi such as *Pythium*, especially in cold, wet soils (plate 166). Fungi that infect aboveground parts of plants can also be combated with pesticides. Fungicides acceptable to organic gardeners include neem oil, a mixture of baking soda and hydrogen peroxide, hot pepper wax, powdered sulfur, lime-sulfur, and copper-containing materials such as Bordeaux mixture and bluestone. Unfortunately, many organic fungicides must be used with great skill and caution to avoid injuring desirable plants. Synthetic chemical fungicides available to home gardeners include mancozeb, chlorothalonil, captan, benomyl, and thiram (all sold under various trade names). Most synthetic fungicides are nontoxic to plants and are much less toxic to humans than are most insecticides, but they may still present hazards both obvious and subtle and must be handled with great care. Any fungicide, whether organic or synthetic, should be applied so as to maximize effectiveness against the target pathogens (thereby reducing the amounts required) while minimizing the impact on nontarget organisms. If fungicides—either synthetic or organic—are used, they should be varied from application to application or at least from season to season. Repeated application of the same fungicide, no matter which one, selects for genetically resistant strains of fungi.

Some fungicides used on ornamental plants such as roses are systemic—that is, the chemicals are absorbed by the plants and cannot be washed off. As a result, any organism that contacts the sap of a systemic fungicide–treated plant will be exposed to the chemical. Although sometimes capable of eradicating already established fungal infections, systemic fungicides are obviously unacceptable for use on plants intended for human or animal consumption or those that pose risks of accidental consumption by pets or children.

Fungicides legally used on edible plants do not enter into plant tissues and are protectant, not eradicant. They can only prevent germination of fungal spores or kill fungi that come in contact with the treated surfaces. Although protectant fungicides cannot eliminate infections already established inside plant tissues, they can reduce their spread from leaf to leaf or plant to plant. Such fungicides must be applied to plant surfaces before exposure to fungal spores, and they must thoroughly coat all exposed surfaces. Applied fungicides can be washed off plant surfaces; degraded by sunlight, heat, and moisture; and metabolized by some bacteria. Protective fungicides (and other pesticides) are often blended with gummy-oily materials known as spreader-stickers to reduce their degradation by ultraviolet light and also their removal by rainfall and sprinkler irrigation. Nevertheless, protective fungicides must often be applied several times in a season to achieve effective disease control.

Despite all efforts by gardeners, fungal diseases can cause severe losses, especially in mild, humid weather or in gardens on well-aerated, warm, sandy soils. Soil-borne wilt pathogens such as *Fusarium* and *Verticillium* pose especially difficult challenges. Crop rotation is rarely effective against soil-dwelling fungal pathogens, because most are capable of surviving on a diet of decaying plant material in the absence of living host plants. Other than soil fumigation, which is rarely feasible for home gardeners, there are no effective chemical controls for such soil-dwelling fungi. Techniques available for home gardeners include soil solarization and growing genetically resistant varieties or nonhost plants.

The gardener battling plant-pathogenic fungi has some tough decisions to make. The most effective control is obtained by continuous coverage of plants with fungicides, but repeated and heavy applications will also maximize exposure of people and beneficial organisms or cause other problems. Thus, whether gardeners consider their practices to be organic or not, it is in their best interests to rely as far as possible on cultural techniques rather than on fungicide applications to minimize fungal diseases. It might be wise to limit fungicides to seed treatments or to emergency situations where conditions are favorable for overwhelming disease epidemics.

Bacteria and Viruses

Beneficial *Rhizobium* bacteria are commercially available for inoculating seeds of beans, clover, alfalfa, and many other legumes (see the mutualism section in chapter 5). Also, some cultures of *Bacillus subtilis* have recently become available for inoculation of seeds and transplants. The best an eco-logically minded gardener can generally do to encourage beneficial bacte-ria is to—you guessed it—enhance the soil organic matter content, avoid soil compaction, and maintain a moderate moisture level.

There are no good chemical controls for pathogenic bacterial infec-tions of plants, although attempts can be made to save especially valuable perennials by treatment with antibiotics such as streptomycin or tetracy-cline (effective against mycoplasmas and mycoplasma-like organisms, MLOs). Prevention or control of bacterial diseases is mostly achieved by cultural practices such as maintaining good garden sanitation; keeping foliage dry; raising resistant varieties when available; intercropping plants that are not susceptible; practicing crop rotation; avoiding spread of infec-tion by contaminated tools, hands, or clothing; and appropriately manag-ing soil pH and fertility. Because most plant-pathogenic bacteria are unable to move significantly under their own power and produce no spores for dispersal by air currents, many bacterial diseases may be controlled by pruning out and destroying (*not* composting) infected plant parts. In addi-tion, many bacteria depend on insect vectors for their spread from plant to plant, and controlling the responsible insects (such as striped or spotted cucumber beetles in the case of bacterial wilt of cucurbits, or leaf hoppers in the case of MLO-caused diseases) may slow or halt the spread of many bacterial diseases.

There are no known cures for viral diseases of plants. Control meas-ures include planting resistant varieties or species, if available; destroying infected plants; rotating crops; planting virus-free seeds and nursery stock; and, most importantly, controlling insect vectors, because most plant viruses depend on insects for their transmission and their survival in the absence of plant hosts. An emerging strategy for controlling virus diseases is vaccinating plants with viruses or other microbes, much as animals are vaccinated, but this technique is not yet well developed. (Although plants lack the biochemical components of animal immune systems, they do exhibit a similar phenomenon of enhanced resistance to infection upon repeated exposure to pathogens. This immunelike response is now being investigated by scientists.)

Managing Plant Nutrients and Soil Amendments

In theory, nutrients could be indefinitely recycled in the garden ecosystem. In practice, however, nutrients are inevitably and continuously lost through leaching to groundwater, runoff and soil erosion, return of nitrogen to the atmosphere via denitrification (discussed in chapter 3), and, most dramatically, through harvesting of plant tissues by gardeners. Harvesting plants from a garden may be considered analogous to mining minerals from deposits in the Earth. If a garden ecosystem is not to become depleted of resources and eventually cease to be productive, materials in a garden must be periodically recharged.

Many gardeners attempt to feed plants by more-or-less routine application of general-purpose fertilizers. However, consideration of ecological factors suggests that this may not be the wisest (nor the most economical) approach. Plants require various amounts of quite a range of nutrients, and not all fertilizers are created equal. Soil and plant tissue testing is available in most areas through commercial testing laboratories or through agricultural extension service offices. Results of such tests may help guide gardeners in deciding whether their plants actually need additional nutrients, and if so, which and how much. Furthermore, observation of plant growth patterns, color of leaves, and so on, can often indicate nutrient deficiencies as accurately as can soil testing, if not more so—it's just slower and more demanding of experience and skilled judgment (plate 167). Additional aid in evaluating nutrient needs can be obtained from extension agents and experienced local growers and gardeners.

Nutrients can be delivered to plants in a variety of ways, not just by applying fertilizers. Soils often contain large reserves of nutrients, such as phosphorus or potassium, in forms unavailable to plants. These reserves may be tapped by manipulating soil pH or by promoting growth of microbes (such as mycorrhizal fungi) that can mobilize these nutrients. Because much usable nitrogen can be extracted from the atmosphere and added to soil by rhizobial bacteria, including properly inoculated legumes in a crop rotation may all but obviate the need to apply nitrogen-containing fertilizers. Compost and organic mulches, which originated as plant tissues themselves, may contain significant amounts of all needed plant nutrients, although in dilute and slowly available forms (plate 168). Green manures or cover crops—plants such as buckwheat, clover, alfalfa, chicory, and others grown simply to be mowed and turned under for soil improvement—can mobilize or add available plant nutrients as well.

Nevertheless, the fastest, simplest, and most common method used by gardeners to administer nutrients is the use of fertilizers. Some of these may be called "organic," others "synthetic," "man-made," or "artificial." All are chemicals, however; they may be beneficial if properly used but are useless or even harmful if misused. For instance, phosphate-containing runoff from overfertilized fields or yards can cause problems in streams and lakes by stimulating excessive growth of algae, leading to a condition in which waters become anaerobic, cloudy, odoriferous, and hostile to desirable organisms such as fish and crustaceans.

Garden fertilizers usually supply the macronutrients nitrogen (N), phosphorus (P), and potassium (K). By law, all materials sold as fertilizer in the United States must state their content of these three macronutrients in percent by weight, hence such notations as 13–13–13 or 7–2–1. Nutrients other than N, P, and K are included in some commercial fertilizers but are stated separately, such as "Ca 10.5" (containing 10.5 percent calcium) or "boron as boric acid, 5 ppm" ("ppm" means "parts per million").

Determining just what sort of plant is to be fed and why will affect choice of fertilizers and rates and timing of application. High-nitrogen fertilizers designed to keep lawns soft and deep green will also produce huge, deep green tomato plants with few, late-ripening tomatoes. Most fruit crops, including tomatoes, produce best with low-nitrogen, high-phosphorus, medium-potassium fertilizers. However, cucurbits and corn/maize require high nitrogen and high phosphorus, probably because of their tremendous sustained growth rates. Fruit trees should be given only low-nitrogen fertilizers and no nitrogen at all after midsummer because trees with succulent tissues from late summer growth spurts are highly susceptible to winter kill. Many annual flowers, such as gladiolus, cosmos, and marigolds, bloom earlier and more profusely and most herb plants are more aromatic if nitrogen is moderately limiting. To green up a lawn or to stimulate lush growth of leafy greens such as cole crops or lettuce, one should apply fertilizer with moderately high nitrogen levels, such as 18–6–6. If trying to reduce transplant shock or to stimulate flowering or fruit set rather than vegetative growth, a fertilizer with a high middle number (P) and much lower end numbers (N and K) would be appropriate, say 10–30–10. Root crops often benefit from extra potassium. Fertilizers with ratings such as 13–13–13 or 12–12–12 are more-or-less balanced blends of all three fertilizer macronutrients and are known as general-purpose fertilizers. Regardless of the particular blend of nutrients, if concentrated chemical fertilizers are used, care must be taken in their application to avoid burning plants by contact with salts, destroying soil structure, lowering the pH of the soil, or contaminating surface waters and groundwater.

Organic fertilizers include mined natural minerals such as phosphate rock, colloidal phosphate (a type of clay), greensand (a source of potassium), flowers of sulfur, granite dust, and saltpeter. They also include by-products of animal and plant processing such as bloodmeal and bonemeal, feather meal, crab and lobster shells, cottonseed meal, and brewers' bottoms. Organic gardeners may also use seaweed (kelp), alfalfa meal, manures, composts, wood ashes, and other materials as plant nutrient sources (plate 169).

Organic fertilizers are much less apt than highly concentrated synthetic ones to burn plants or cause environmental contamination and are also less prone to acidify or salinize soil. They may add beneficial organic matter and micronutrients as well as macronutrients and may release their nutrients over a much longer period. However, even organic fertilizers have potential problems. They tend to be bulky and low in nutrient concentrations, and large quantities must be mined or generated, packaged, hauled, and applied—processes that in themselves cause environmental problems such as high consumption of fossil fuels or contamination of groundwater by mining operations. Organic fertilizers also tend to be expensive and their nutrients too slowly released for plants under some conditions. And, although generally less concentrated than synthetic fertilizers, they still are not entirely safe. Manures can burn plants and contaminate water, and there are concerns about possible pathogenic viruses and other microorganisms in manure, blood, bonemeal, and other animal-derived materials. Particular concern has been expressed over possible contamination of bloodmeal (obtained as a by-product of the slaughter of cattle), a popular organic source of nitrogen, with the prions that cause fatal mad cow disease and its equally deadly human version, Creutzfeldt-Jakob syndrome.

Soil amendment is a catch-all term for substances added to soil that do not fulfill the legal definition of fertilizer. These include not only widely used lime and gypsum, but also a strange and wonderful array of other materials, including every sort of ground mineral or rock, detergents, glues, extracts of seaweed, and diatomaceous earth. These substances can range from beneficial to toxic. The general rule of thumb is *caveat applicator*, "Let the applier beware." Wise gardeners try out an unfamiliar product on a small scale before using it on an entire garden or yard, no matter how ecstatic the testimonials published in the sales literature.

Managing Energy

Major goals of ecologically responsible gardening include maximizing internal recycling of nutrients and other materials; minimizing inputs, particularly those obtained from distant sources or produced by energy-intensive or polluting industrial processes; and reducing outputs, especially of energy-rich or toxic materials. Energy cannot be recycled, however, and gardens always suffer losses of energy and require inputs. Energy losses from a garden and the consequent need for replenishment may be especially substantial if nutrients and energy are removed in the course of harvesting vegetables, fruits, flowers, or even weeds. However, the efficiency with which energy is used can vary enormously with differences in gardening practices.

Some gardening tactics can reduce the need for energy inputs. Choose a site as close as possible to where tools are stored and harvested crops are used. Avoid steep slopes that require energy-consuming civil engineering to prevent erosion or necessitate up-and-down-hill travel. Raise plants well-adapted to the local climate and soil. Practice water conservation to reduce the need for irrigation. Return as many plant residues as possible to the soil. And plant nitrogen-fixing legumes to minimize the need for energy-expensive nitrogen fertilizers.

Gardening is essentially impossible without least some inputs of energy. The ideal energy source for a garden is the sun—clean, silent, and free. Thus, anything that can reasonably be done to enhance use of solar power is likely to be advantageous. Gardeners can strongly influence the efficiency with which sunlight is used by appropriately choosing and positioning plants; being attentive to plants' particular shading and exposure requirements; applying mulches; and placing cloches and other protective structures.

Plants such as cacti and other succulents, corn/maize, and sorghum use bright sunlight more effectively than most others do. Positioning a garden to escape shade cast by tall structures or nearby plants often greatly enhances productivity of these plants. Shade plants such as hostas, trilliums, and many houseplants grow best with reduced light intensity. Matching garden plants with their appropriate environments will optimize use of energy; to do otherwise is a waste of resources, at least to some degree.

Mulches come in a wide range of colors and materials and have varying effects on plants. Organic mulches such as straw and compost reduce light absorbed by the soil and thus cool it. Paper mulches are available in a range of colors and weights and may either modestly warm the soil (dark colors) or cool it (white). Unfortunately, these mulches, while biodegradable and made of renewable resources, may create as much pollution in their manufacture and distribution as plastic mulches—and generally weigh and cost much more. For small gardens, though, an excellent and widely available organic mulch is black-and-white newsprint (*not* colored—many color inks are toxic to plants). Old newspapers can be laid down like overlapping shingles or taped together to make strips. However, they should be applied at least five sheets thick or weeds can push through them. Newsprint will completely biodegrade into the soil and not have to be removed. Unfortunately, no organic mulches have the strongly soil-warming capabilities of clear or black plastic sheeting, nor are they as convenient to ship, store, or apply.

Plastic mulches (other than white or silver ones) generally warm soil. The most effective at warming soil is clear plastic sheeting. On sunny summer days it may even heat the soil too much for plant roots to tolerate, but it can kill many harmful soil organisms. One drawback of clear plastic mulch in spring and autumn is that warming the soil at these times benefits weeds as well as desirable plants. Black plastic mulch effectively prevents weed growth by blocking out sunlight, but the warming effect is not nearly as strong as that of clear plastic. Recently developed infrared-transmitting (IRT) plastic mulches, usually brown or green in color, transmit soil-warming infrared sunlight while absorbing or reflecting the wave-lengths of light needed for photosynthesis, thus combining the beneficial properties of both clear and black plastic. The main drawback of IRT film is its significantly greater cost. Other specialty mulches include red plastics put under tomato and other fruiting plants to enhance fruit production and metallic foils that reflect almost all light, illuminating plants from beneath to enhance photosynthesis by lower leaves. One significant prob-lem with virtually all plastic mulches is the environmental difficulties involved in their disposal after use. Just how does one dispose wisely of large pieces of tattered polyethylene sheeting?

Floating row covers, hotcaps, cloches, and polyethylene tunnels may be used alone or in combination with mulches to act as miniature green-houses, trapping and converting incoming sunlight into heat energy and conserving heat energy in the ground by preventing its loss into the atmos-phere. In cases where incoming sunlight is too intense, shade cloth, usually black or green nonwoven plastic mesh, may be hung over plants.

The angle of incoming light strongly influences its effectiveness in warming the garden and driving photosynthesis. Because of this, garden-ers can take advantage of slope-generated microclimates to successfully raise plants in climate zones normally too cold or hot for them. This effect is most strongly seen in winter on ground that slopes toward or away from the equator. A rough rule of thumb for the Northern Hemisphere is that each degree of slope to the north or south alters solar heating of a piece of ground equivalent to moving it 30–50 miles (50–80 km) away from or closer to the equator, respectively. The effects of northerly and southerly slopes are reversed in the Southern Hemisphere, of course. The practical consequence of the slope-generated differences in sun angles is difference in soil temperatures. The near-surface temperature of a southerly slope may be as much as 12°F (7°C) warmer on a sunny winter day than a neigh-boring northerly slope of similar characteristics. Thus, an enterprising gar-dener can create raised east–west ridges of soil in the garden and grow plants only feet apart in solar radiation microclimates similar to those of level ground hundreds of miles apart (plate 170). Because of heat transfer by conduction through air and soil and sunlight directly striking the plants themselves, the total difference in microclimate might not actually be so dramatic—but there would be a substantial difference nonetheless.

In areas where the air tends to be chilly, especially at night, gardeners can take advantage of the relatively great heat capacity of rocks by growing plants along the south faces of stone or masonry walls (a common gardening practice in England and other parts of Europe) or by surrounding plants in small depressions with large rocks (as do farmers in the frigid highlands of Bolivia, Peru, and Tibet). During the day, the rocks soak up energy from solar radiation, then gradually release it at night as heat, warming adjacent plants in the process (plate 171).

The fact that colder air is denser than warmer air creates another microclimate effect that may affect gardens in hilly or sloping areas. Cold, dense air tends to flow downward into valleys, displacing lighter, warmer air, which then rises toward the hilltops. Generally, valley bottoms have cooler nights, later spring frosts, and earlier autumn frosts than do nearby upper hillsides or hilltops. Even a difference in elevation of 200 feet (60 m) from top to bottom of one of the hills on Steve Salt's farm makes a difference in air temperature of as much as 10–15°F (6–9°C) on calm, clear, spring and autumn nights—and a difference of more than a month in the frost-free growing season. Frost-sensitive plants benefit in spring and autumn from the warmer hilltop microclimate, whereas heat-intolerant plants such as lettuce, spinach, larkspurs, and delphiniums enjoy the cooler summer nights in the frosty bottoms.

Another aspect of garden energy management, perhaps the one most immediately apparent to many gardeners, is the use of tools and equipment. All tools consume energy to do work, some more efficiently than others. Animal engines (including humans) are generally more fuel-efficient than most internal combustion engines. Diesel engines are usually more efficient than gasoline-powered engines; and four-cycle gasoline engines more efficient than two-cycle ones. And all engines operate more efficiently and emit fewer noxious by-products if they are well maintained and tuned up (plate 172).

Many gardeners cringe at the thought of the blisters, sweat, and sore backs associated with use of hand tools. However, good-quality hand tools, when well maintained and properly used, can be surprisingly productive and easy to use. Purchase good-quality tools: the real difference between a cheap tool and a good one may not be obvious at the hardware store, but it is often painfully evident after several hours or weeks of garden work. Keep tools well maintained: having all cutting edges sharpened, metal surfaces cleaned and oiled, and bolts and nuts tightened makes a world of difference in how a tool performs and lasts (plate 173). Use tools properly: there is as much technique involved in skillful use of garden hand tools as there is in the use of sports equipment or carpentry tools; unfortunately, there are few instructors available nowadays to teach hand tool techniques that were once common knowledge on homesteads and farms.

So how *can* a gardener learn old-time garden hand tool skills? Good question. You may be lucky enough to locate in libraries or antiquarian bookstores late-nineteenth-century texts on how to farmstead, or, even more fortuitously, to make contact with a rare old-timer who can teach some of these techniques. However, what survives of this knowledge generally now seems scattered through high-end garden tool catalogs, books on gardening methods, and an occasional article in gardening magazines. Thus, it is likely that you will have to systematically experiment on your own to find out, for instance, which ways of gripping tool handles or which angles of motion are most efficient and least tiring.

If a job is infrequent or minor, a gardener may consider using hand tools or renting a power tool. If a garden task is frequent, important, and beyond the reasonable capabilities of hand tools, then an appropriately sized power tool may be called for. As with hand tools, skillfully using good-quality, well-maintained power tools that are well matched to the jobs at hand will go a long way toward maximizing their benefits in gardening, reducing stress on the gardener, and minimizing undesirable effects on the environment.

One common question is whether to use specialized tools that do one or a few closely related jobs, but do them well—as opposed to more generalized equipment capable of doing many things, but possibly none of them well. For example, a garden tractor may be capable of powering a wide range of devices from mower decks to sprayers, but may operate none of them as well as specialized devices dedicated to mowing, spraying, between-row weeding, compost-turning, or so forth. And even in the categories of sprayers and applicators, specialized models may do excellent jobs of applying particular types of pesticides, fertilizers, or other substances, yet be totally unsuited for others. Only a manure spreader can do a good job of spreading manure. A lawn mower would soon break if called upon to cut woody brush that presents no challenge to a brush mower—but a brush mower would leave a lawn terribly ragged. There is no universal answer to this question, but the wise gardener will consider the global costs and benefits, comparing the financial and environmental costs of purchasing and maintaining several pieces of specialized equipment versus the possibly greater fuel consumption, time requirements, personal effort, and repair costs of using generalized tools marginally suited to the tasks at hand. Durability of equipment has consequences—replacing or repairing a tool may impose greater environmental costs than operating a slightly less fuel-efficient, but significantly more durable one made of heavier materials. Also to be considered are heat, pollutant, and noise output; spillage and disposal of fuel and lubricants; and consumption of power by such things as excessive idling or "bells and whistles" like power steering and electric starters instead of pull cords.

Holistic Garden Management

Gardeners are faced with numerous difficult decisions: choosing which plants to grow, encouraging beneficial organisms and discouraging pests, caring for the soil and water, and so forth. Furthermore, even beneficial gardening activities often seem to compete for common resources or conflict with one another. Thus, it is important that a gardener evaluate the environmental impact of gardening practices holistically and globally. Holistic analysis means that all costs and benefits of practices and equipment should be taken into account, not just the immediately apparent aspects. For example, a gardener should consider the ultimate impacts of the production, packaging, transportation, application, use, and final disposal of all tools, equipment, and material used. Global analysis means that environmental costs or benefits that are remote to the garden and gardener in time or space should be identified and considered. For instance, the costs of obtaining raw materials and manufacturing a piece of equipment or of supplying fuel or electric power may be remote to a particular garden, but they are just as consequential as are more immediate and obvious fuel consumption, noise, and local pollutant output. Out of sight, out of mind—but not out of existence! Failure to think and act both holistically and globally may result in a gardener (or anyone else, for that matter) short-sightedly adopting apparently good practices that are actually more harmful than others.

For example, a gardener might decide to replace a gasoline-powered piece of machinery with an electrical one with the goal of reducing the environmental impact. However, it should not be forgotten that the power plant generating the electricity might burn fossil fuels and release pollutants and that there are usually great losses of energy during long-distance transmission of electricity. Also to be considered are substantial inefficiencies both in the generation of electricity and in its conversion into mechanical power. It is possible that a clean, quiet, electrical machine won't look so much better than a noisy, polluting, gasoline-powered one after a global and holistic analysis of all factors. Of course, human sweat-powered machines are much more energy efficient than any engine-powered ones, and the fuel that they burn may be potentially life-threatening fat deposits. So, a gardener might ultimately decide to use a hand tool instead of an engine-powered one and work out in the garden instead of at the health club.

Other cost-benefit analyses may focus on the extent of use (or nonuse) of pesticides and fertilizers. All substances applied in the garden—including organic ones—impose substantial environmental costs in their production, transportation, distribution, use, and disposal, yet few gardeners and virtually no farmers are willing to forswear their use. The ecologically astute gardener or farmer will, however, weigh the costs and benefits of all alternatives for pest control and plant nutrition and make decisions that optimize the trade-off between environmental costs and economic or aesthetic benefits.

Aesthetic benefits may impose other costs as well. No responsible person would knowingly turn loose a plague in his or her neighborhood, yet many gardeners frequently risk disrupting local ecosystems by planting beautiful but potentially invasive exotic ornamentals. Purple loosestrife entered this country as an ornamental and still is a beautiful . . . plague. At the least, a wise gardener should seek information about the biological characteristics of a candidate garden plant that might make it an aggressive weed, such as spread by underground runners or rhizomes, production of wind-blown or bird-carried seeds, prolific self-reseeding, and so forth (see chapter 4). This is not to say that all—or even most—exotic plants are environmentally hazardous, but an ecologically minded gardener would certainly want to identify those that likely are and avoid them, or at least take pains to prevent their spread.

And what of growing plants so poorly adapted to the local environment that they require expensive and environmentally disruptive gardening techniques such as heavy irrigation, substantial pesticide applications, soil acidification, or extensive protection from the local climate? One might hope that such plants would be raised only if they provide some substantive benefit to offset their environmental costs.

In all of these cost-benefit analyses, there is no black or white but shades of gray. The natural world is a complex and interdependent realm with few simple answers. For instance, planting an orchard on a steep slope rather than on level ground may furnish benefits in the way of ample sunlight exposure, good air and water drainage, reduced frost damage, decreased disease and pest damage to the trees (thus less need for pesticides)—and healthy exercise for the orchardist. But this decision may also result in increased fuel consumption, aggravated air and water pollution, greater wear-and-tear on equipment (and thus greater costs associated with maintenance, repair, and replacement), and potentially serious soil erosion. Increasing plant spacing in the garden will reduce pest and disease problems, decrease competition by plants for water and nutrients, and facilitate harvesting and other garden operations, but it will also enhance germination and growth of weeds, increase wind and sunburn damage, and possibly increase soil erosion. Closer spacing of plants will have different pros and cons.

So what are the right choices? Decision-making really depends on the quality of information available to the decision-maker, and even more significantly, on his or her personal ethical value system. Also, it is rare that gardeners (or anyone else, for that matter) have complete freedom in pursuing the options available to them. Limitations of time, space, money, interpersonal relationships, and so forth may constrain them to make less than optimal choices.

The best one can do is to seek out as much relevant and reliable information as possible and analyze a proposed course of action both holistically and globally and in light of one's personal ethical value system—then proceed with diligence and prudence, stopping frequently to examine and analyze the consequences of one's actions. Perfect? No, but it's a great deal better than acting blindly. In this book we've tried to reveal some of the myriad interactions and actors in the drama of garden ecology, many of them not obvious, perhaps even counter-intuitive. What we do with this knowledge is now up to each of us.

Epilogue

In this book, we have placed gardening in a broad ecological context. To do this, we've looked beyond the garden border and contemplated ways in which gardens and garden organisms influence and are influenced by the world in which they are found. We hope the message that emerges is that gardens are not islands, isolated from the natural world (plate 174). Rather, by virtue of their connectedness through the atmosphere, soil, water, and movement of organisms, gardens are as integral a part of the Earth's ecosystems as are its tropical forests, coastal wetlands, and prairies. Gardeners who keep this in mind are less likely to disrupt these ecosystems and more likely to contribute to the improved health of our planet.

Gardening based on ecological principles should result not only in healthier plants, but also in decreased stress on local and regional environments. By understanding the garden environment and the ecological requirements of garden organisms, ecologist-gardeners can make decisions about plant choice and care, soil management, and dozens of other gardening practices on the basis of knowledge and judgment rather than convention and habit. For example, a gardener who knows that a suddenly common insect will not significantly damage plants need not invest limited time, money, or other resources in an attempt to control it, nor alter the habitat with unnecessary and possibly ineffective chemicals.

Ecologically sound gardening can benefit the environment well beyond the garden and yard. Over the course of a growing season, we make dozens—perhaps hundreds—of decisions that can have far-reaching effects. These decisions, involving issues from material and energy costs of our purchases and practices to whether to apply pesticides or fertilizers, may have global implications. For example, gardeners who take care of their tools not only work more efficiently, but also conserve the resources used in manufacturing and marketing these items because they are likely to last longer. And the gardener who abstains from spraying a relatively harmless insect conserves resources and may safeguard surface water, groundwater, soil, and nontarget organisms.

It is estimated that half the households in the United States have a flower or vegetable garden. This represents a potentially loud voice in the interests of planet Earth. Consider the result if each of us recycled the packaging of the gardening supplies we purchase rather than throwing them in the trash, avoided harming—or better yet, encouraged—the native insect pollinators and predators in our gardens, and composted kitchen and yard waste to improve our soil. These simple acts, multiplied tens of millions of times, would likely have a profound effect on problems such as solid waste. Gardening can also improve the quality of the environment by increasing soil organic matter and decreasing soil erosion; providing habitat and food sources for native pollinators, songbirds, and other wildlife; serving as a natural filtration system for household gray water; reducing energy requirements for heating and cooling buildings; providing locally grown food and ornamental plants that require little, if any, packaging, refrigeration, and shipping; preserving biological diversity; increasing the amount of green space in a world prone to development; and much more.

Gardening satisfies many needs, and those of us who garden do so for many reasons. Growing plants in an ecologically sound manner is likely to produce healthy plants as well as contribute to our physical, emotional, and spiritual well-being. Gardening provides a means of exercise and an opportunity for meditation and reflection, thereby improving our physical and mental health. Gardening brings us outdoors, gets our hands in the earth, and takes us away from our usual routines. It helps us shed the frustrations of the day as we focus on the beauty and life in front of us, but also keeps us in touch with the sometimes harsh realities of the natural world. Gardening can be a private activity or an opportunity to share the day's events with family, friends, and neighbors while transplanting, weeding, and harvesting. Gardening can provide us with fresh, nutritious food and with types and varieties of fruits, vegetables, and cut flowers that may not be locally available. Gardeners with sufficient space can donate food to soup kitchens and other venues that feed those who are less able to provide for themselves; they can share fruits, vegetables, and flowers with family and friends; or they can become market-gardeners, supporting local economies by offering for sale fresh, home-grown produce and flowers at farmers' markets or elsewhere.

Approaching gardening from an ecological perspective doesn't require a lot of land. Although many gardeners tend plots in their own yards, the benefits of gardening are not limited to those who own property or to those who can convince their landlords to give up a piece of the lawn. Containers can bring many of the pleasures and benefits of gardening to the patio, deck, or urban balcony. Many cities have community gardens where space can be rented, and more and more cities have downtown beautification and adopt-a-garden programs that rely on volunteers for care of their plantings. Gardeners wishing to learn from others can take classes and workshops at area schools and garden centers or through extension services. In these and other ways, those interested in gardening can get their hands into the soil.

Although gardens are created environments, they can be thought of as windows into the workings of the natural world. To the extent that we approach gardening with the welfare of the natural world in mind, our gardens will be places in which to nourish plants and ourselves and through which to make Earth a healthier, better place in which to live.

Glossary

Words italicized within definitions are themselves defined in the glossary.

abscission: loss of a part, such as a leaf or *fruit*, due to development of a weakness in the point of attachment

adventitious root: root that forms at an uncharacteristic location, such as along a stem

aerobic: referring to conditions, such as in soil or water, in which oxygen is present or to organisms that require oxygen for *cellular respiration*

after-ripening: physiological changes that must occur in certain dormant *seeds* for *germination* to take place

aggregate: in relation to soil, a mass of particles held together by clay, adhesive molecules, and organic matter; aggregates help impart structure to soil

air layering: means of propagation in which a stem is wounded, treated with plant growth *hormone*, and wrapped in peat moss or similar material; when *adventitious roots* form at the wound, the section of stem above the wound can be removed and planted

allelopathy: release of a chemical by one plant that interferes with growth or reproduction of another nearby plant

alternation of generations: reproductive life cycle of plants that involves two alternating, multicellular phases, or generations; the transitions between these phases involve either the halving of *chromosome* number (during sperm and egg formation) or the doubling of chromosome number (during the union of sperm and egg)

anaerobic: referring to conditions, such as in soil or water, in which oxygen is not present or to organisms, such as certain bacteria, that do not require oxygen

angiosperm: a plant that produces true *flowers* and whose *seeds* are enclosed within a *fruit*

annual: a plant that is able to complete its life cycle within one *growing season*

anther: usually terminal, inflated portion of the *stamen* in which *pollen grains* are produced

antibiotic: substance produced by one organism that is toxic to another

apical dominance: tendency for buds lower on the stem to be suppressed by *hormones* transported from terminal (apical) buds

aposematic coloration: conspicuous coloration that signals an organism is distasteful, toxic, or dangerous in some way

axil: angle formed between a branch or leaf petiole and the stem

biennial: a plant that requires two years to complete its life cycle; during the first year, *germination* of the *seed* and vegetative growth occur, with reproduction the following year

biodiversity: total number of species found in an area

biological control: use of predators, parasites, herbivores, or other living organisms in an attempt to control a population of another species

bolting: rapid increase in height of a rosette plant prior to flowering in response to *hormones* or stress

carnivory: interaction in which one organism eats the flesh of another, especially applied to animals that eat other animals

carpel: female, or *ovule*-producing, part of the *flower*, generally including a *stigma*, *style*, and *ovary*; used interchangeably with *pistil*

cellular respiration: series of chemical reactions in which glucose or other carbohydrates are broken down to yield energy, carbon dioxide, and water

chelator: molecule that can reversibly bind and release other, usually electrically charged, molecules

chlorophyll: green plant pigment that absorbs light during *photosynthesis*

chloroplast: cell *organelle* that contains *chlorophyll* and is the site of *photosynthesis* in plants

chlorotic: plant tissue that is yellowed, resulting from the breakdown or poor development of *chlorophyll*

chromosome: long structure within a cell, comprised primarily of DNA, that contains the genes

cleistogamy: *self fertilization* in a usually highly reduced *flower* that does not open

climax community: *community* that is expected as the final, self-perpetuating stage of *succession* for a given area

commensalism: interaction between individuals in which one benefits and the other neither benefits nor suffers harm

community: all the organisms that exist together in a habitat

companion plant: species that, when planted in close proximity to another plant, seems to benefit the second species by repelling harmful organisms, attracting beneficial organisms, or by some other means

competition: interaction in which individuals reduce the amount or availability of a limited resource required by others; by definition, all individuals involved in a competitive interaction suffer a net disadvantage

constitutive defense: plant defense, such as a chemical compound effective against herbivores, that is present in tissues at all times

cotyledon: structure in *seeds* that either stores or absorbs nutrients; seeds of flowering plants may have either one or two cotyledons (in monocots and dicots, respectively)

cover crop: crop that is planted, allowed to overwinter, and then generally turned under the following spring; planted to decrease winter erosion, increase soil nutrient levels, or improve the soil in some other way

crop rotation: growing different successive crops in a garden or field; often done to deny pests the same food source as nourished their population the previous year or to balance use of soil nutrients

cross-fertilization: *fertilization* in which the egg and sperm are produced by different individuals

crypsis: characteristic of an organism, such as its coloration, that allows it to blend into its background as a means of avoiding detection

cultivar: variety of a plant that has been bred for cultivation; term coined from "cultivated variety"

cuticle: waxy layer over the *epidermis* that reduces water loss from structures such as leaves

day-neutral plant: a plant whose flowering is not triggered by changes in day length

degree-days: unit of heat in which the number of degrees above a crop-specific temperature threshold are determined daily and summed over a period of time; used in predicting stages in development such as time of flowering and *fruit* maturity

detritivore: organism that derives its nutrition from breaking down tissues of dead plants, animals, or microbes

dominant species: a species that is of critical importance in a *community* by virtue of its numbers, biomass, or functional role

double fertilization: key element in the reproductive process of flowering plants involving both sperm cells contained in the *pollen grain*: one sperm cell combines with the egg to form the *embryo*, whereas the other contributes to formation of the *endosperm*

ecology: the investigation of relationships among organisms and between organisms and their environment

ecotone: transition area where two adjacent *communities* intermingle

edge effects: various ways in which organisms and their populations respond at the edges of habitats (including gardens) in comparison to their behavior toward the centers of the same habitats

embryo: part of the *seed* that results directly from *fertilization* of the egg by the sperm and that develops into the *seedling*

endosperm: nutritive tissue in the *seeds* of flowering plants; one of the products of *double fertilization*

epidermis: outermost layer of cells in leaves, *flowers*, and most other plant parts; usually one cell thick

epiphyte: plant that grows on another plant or that uses another plant for physical support but that is not parasitic

etiolation: growth condition involving excessive stem elongation, pale color, and other symptoms, usually due to inadequate light

eukaryote: an organism, including members of the plant, animal, fungal, and protist kingdoms, whose cells include *nuclei*, *mitochondria*, and other characteristic structures

exoskeleton: external, protective armor comprised primarily of chitin; characteristic of insects and their relatives

extrafloral nectary: nectar gland located outside of a *flower*

facultative parasite: parasite that is capable of existing in a free-living state in the absence of a living *host*

fallow: to leave a field or garden plot unplanted as a means of decreasing the number of garden pests in the soil or to accomplish another goal such as water conservation

fertilization: union of a sperm and an egg; in flowering plants, this occurs in the *ovule*

fibrous roots: root system consisting of many roots of approximately equal size

filament: stalked portion of the *stamen* supporting the *anther*

floating row cover: lightweight, nonwoven polyester or polypropylene cloth that is placed over plants to exclude pests or retain heat

flower: reproductive structure in *angiosperms* that includes the *pollen* and/or *ovule*-producing organs

food web: the sum of all links among organisms in an area, each link representing energy flow resulting from feeding

frass: the feces of borers or other herbivorous insects; a crumbly material that resembles sawdust, it may be more easily detected than the insects themselves

fruit: mature, ripened *ovary* and sometimes associated structures of a plant; most fruits form only after *fertilization* of at least some of the *ovules* within

gall: abnormal growth on a plant caused by the presence of an insect, bacterium, or other organism, often as a result of the production of a plant *hormone* or hormone mimic

genus (pl. genera): taxonomic category containing closely related species

geotropism: tendency of a plant or plant part to grow toward (positive) or away from (negative) the force of gravity

germination: beginning of growth in a *seed* in which the *radicle* emerges from the *seed coat*

grafting: means of propagation in which a leaf, branch, or stem of one plant is fused to another plant

green manure: crop that is planted in the spring and turned under during the same *growing season*; used to increase nutrient levels or improve the soil in some other way

growing season: number of days between the average date of the last killing frost in the spring and the average date of the first killing frost in autumn

half-hardy: horticultural term for plants that are able to survive (but are often injured by) a brief killing frost (28°F, −2°C), but that generally cannot survive through a winter

hardy: horticultural term for plants that are able to survive and grow at temperatures below 28°F (−2°C) and may survive sustained winter weather

haustoria: specialized *hyphae* (in parasitic fungi) or roots (in parasitic plants) that penetrate their *hosts* and withdraw nutrients and water

heat-units: see *degree-days*

herbivory: consumption of part or all of a plant, usually by an animal

horizon: in relation to soil, a horizontal layer that can be distinguished from other such layers on the basis of texture, color, and chemistry

hormone: substance produced in small amounts in one part of a living organism and transported to sites where it controls growth and development processes

host: organism that provides food, shelter, or other benefit to an individual of another species, for example, a parasite or herbivore

humus: usually dark, decay-resistant fraction of the soil organic matter; responsible for much of the water- and nutrient-absorption capabilities of soil

hyperparasite: organism that parasitizes another parasite; for example, a fungus that parasitizes a plant-parasitic fungus

hypha (pl. hyphae): tubular, threadlike filamentous tissues of a fungus

induced defense: plant defense, such as a chemical that is effective against herbivores or parasites, that is produced in response to the *herbivory* or *parasitism* itself, but is normally absent otherwise

infection: establishment of a foreign organism, such as a parasite, in its *host*

inoculate: to introduce microbes into an environment where they may become established, as in the movement of *Rhizobium* bacteria into a garden planted with legumes; this may occur naturally or by human action

instar: stage in the life cycle of an insect between two successive molts

keystone species: species whose ecological role is critical to the structure of its *community* and whose removal would result in dramatic change

layering: means of propagation in which a branch or stem is laid against the ground surface and covered to induce rooting

light compensation point: for a particular plant, the amount of light at which the rate of formation of sugars by *photosynthesis* is equal to the rate of breakdown of sugars by *cellular respiration*

light saturation point: for a particular plant, the amount of light above which the rate of *photosynthesis* does not increase

limiting factor: resource whose availability is less than that required by an individual and therefore restricts growth, even if all other required nutrients are present in abundance

long-day plant: a plant that is stimulated to flower when the number of hours of daylight exceeds a critical value, as in the spring; the actual stimulus, however, is a critically short period of darkness

macronutrient: chemical element, such as nitrogen or phosphorus, that is required by plants in relatively large amounts

marcottage: see *air layering*

meristem: plant tissue whose cells retain the ability to continue dividing

mesonutrient: chemical element, such as calcium or sulfur, that is required by plants in intermediate amounts

metamorphosis: series of changes through which insects pass in developing from egg to reproductively mature adult; complete metamorphosis includes egg, larva, pupa, and adult stages; incomplete metamorphosis involves changes in which successive stages (*instars*) look increasingly like the adult

micronutrient: chemical element, such as iron or zinc, that is required by plants in relatively tiny amounts

mimicry: aspect of appearance or behavior in which an individual resembles an inanimate object or a member of a different species, often as a means of deceiving predators or prey

mitochondrion (pl. mitochondria): cell *organelle* in which *cellular respiration* takes place

mutualism: interaction involving individuals of different species in which all benefit

mycelium: mass of *hyphae* that together makes up the body of a fungus

mycorrhiza (pl. mycorrhizae): *mutualism* involving certain species of fungi and the roots of plants

node: area of a stem where leaves attach

nodule: swelling on roots resulting from a *mutualism* between nitrogen-fixing bacteria and certain plants, especially legumes

nucleus: cell *organelle* that contains the *chromosomes* and that directs most cell activities

obligate parasite: a parasite that cannot live independent of its *host*; often restricted to one or a very few species or even strains of hosts

organelle: membrane-enclosed structure, such as a *chloroplast*, *mitochondrion*, or *nucleus*, having a specific function within a cell

outcrossing: see *cross-fertilization*

ovary: basal portion of the *carpel* that contains the *ovules* and develops into a *fruit* following *fertilization*

ovule: multicellular structure that develops into a *seed* when mature, usually following *fertilization*; in flowering plants, ovules are enclosed within the *ovary*

parasitism: interaction in which one individual (a parasite) removes nutrients, fluids, or tissues from another individual (a *host*), often doing so over an extended period of time and usually without directly killing the host

parasitoid: specialized parasite that generally kills its *host*; most commonly, wasp and fly larvae that feed on the soft tissues of hosts

parthenogenesis: form of asexual reproduction in which genetically identical offspring (clones) are produced

perennial: plant that has the potential to live many years

petal: often colored and leaflike floral part located just inside the *sepals*; often functions in attracting pollinators

petiole: leaf stalk

phenology: timing of events such as flowering and *seed* maturation

pheromone: chemical produced by an animal to communicate with others of the same species, for example, for the purposes of mating

phloem: the food-conducting tissue of plants

photorespiration: stimulation of *cellular respiration* by intense light; may cause unproductive depletion of plant sugars

photosynthesis: series of chemical reactions in which carbon dioxide and water are combined to produce glucose and oxygen; depends on the absorption of light energy by plant pigments, especially *chlorophyll*

phototropism: tendency of a plant or plant part to grow toward (positive) or away from (negative) a light source

pistil: see *carpel*

pollen grain: tiny, decay-resistant structure formed in the *anther* that contains two sperm cells at maturity

pollen tube: structure that forms after germination of a *pollen grain* and through which sperm are delivered to the *ovule*

pollination: transfer of *pollen grains* from an *anther* to an appropriate *stigma*; the anther and stigma may be part of the same *flower*, in different flowers on the same plant, or in flowers on different plants

predation: interaction between individuals in which one is eaten by the other; in classical, or carnivorous, predation, an organism other than a plant or fungus is consumed

prokaryote: an organism, including members of the eubacteria and archaea kingdoms, whose cells lack *nuclei* and most *organelles* characteristic of *eukaryotes*

propagule: any structure, such as a *seed*, spore, or appropriate vegetative part, that can disperse and establish a new individual

radicle: part of the plant *embryo* that develops into the root and the first structure to emerge from a germinating *seed*

reservoir: in relation to *parasitism*, a pool of potentially infectious organisms that can infect suitable *hosts* should conditions become favorable

rhizome: a stem adapted for vegetative spread that grows horizontally below the surface of the ground

rhizosphere: volume of soil immediately surrounding the roots

root hair: tiny projection from the wall of a root cell through which water and dissolved nutrients are taken up

scarification: weakening or cutting of the *seed coat* as a means of increasing the likelihood of *germination*

seed: fertilized *ovule*, normally containing an *embryo* and nutritive tissue (the *endosperm*)

seed bank: *seeds* accumulated in the soil that were produced during previous years and are still viable

seed coat: protective outer layer of a *seed*

seedling: young plant, shortly after seed *germination*

self-compatible: able to undergo *self-fertilization*

self-fertilization: *fertilization* in which both the egg and sperm are produced by the same individual

sepal: outermost *flower* part that encloses and protects the other flower parts

shade cloth: nonwoven, often plastic, mesh used to cover plants to decrease the intensity of sunlight

short-day plant: a plant that is stimulated to flower when the number of hours of daylight declines below a critical value, as in late summer or autumn; the actual stimulus, however, is a critically long period of darkness

solarization: heating soil using solar energy, for example, by covering the ground with clear plastic

specific epithet: second part of the Latin name of a species, which, when combined with the *genus*, provides an unambiguous, unique name for an organism

spiracle: small pore in the skin or *exoskeleton* of an insect through which oxygen, carbon dioxide, and other gases are exchanged

stamen: *flower* part having a pollen-producing *anther*; frequently borne at the end of a narrow *filament*

steward: in the context of gardening, one who manages the land and its inhabitants in such a way that they are not degraded, either in the short term or the long term

stigma: portion of the *carpel* on which *pollen grains* are deposited during *pollination*

stolon: stem adapted for vegetative spread that grows horizontally across the surface of the ground

stomate: microscopic pore, especially common on leaves, through which water vapor and other gases pass

stooling: see *layering*

stratification: subjecting *seeds* to sufficiently low temperatures for an extended time period, often weeks, to overcome barriers to *germination*

style: narrow portion of the *carpel* that connects the *stigma* and the *ovary*

stylet: piercing mouthpart in invertebrates such as nematodes or aphids

succession: process by which earlier established species are displaced by later colonizing species

systematics: classification and study of organisms with regard to their natural relationships

taproot: main root in a system in which one root is significantly larger than the others; forms directly from the *radicle*

tender: horticultural term for plants that may survive, but that are generally injured by, brief exposure to a light frost (32°F, 0°C) and normally cannot survive a killing frost (28°F, −2°C)

transpiration: loss of water vapor from a plant due to evaporation, especially through *stomates*

trap crop: species planted in a garden to lure insects or other pests away from more desirable or valuable plants

vacuole: membrane-enclosed, fluid-filled volume that can occupy 90 percent or more of a plant cell; provides internal pressure and stores pigments

vector: in relation to *parasitism*, an agent (such as the wind, an insect, or a garden tool) that transmits a disease-causing organism to a potential *host*

vein: an often-prominent strand of fluid-conducting *xylem* and *phloem* in a leaf

vernalization: process in which flowering is induced by exposure to a minimum period of cold

very tender: horticultural term for plants that are killed or severely injured by above-freezing chilling or even brief exposure to a light frost (32°F, 0°C)

xeriscaping: cultivating plants adapted to dry conditions as a means of conserving water

xylem: plant conductive tissue that transports water and dissolved nutrients; also has support and storage functions

zygote: first cell resulting from *fertilization*; combines the genes of the sperm and the egg

Further Reading

Brickell, Christopher, and Elvin McDonald, eds. 1993. *The American Horticultural Society Encyclopedia of Gardening*. New York: Dorling Kindersley.

Brickell, Christopher, and Judith D. Zuk, eds. 1996. *The American Horticultural Society A–Z Encyclopedia of Garden Plants*. New York: Dorling Kindersley.

Capon, Brian. 1990. *Botany for Gardeners*. Portland, Oregon: Timber Press.

Damrosch, Barbara. 1988. *The Garden Primer*. New York: Workman Publishing.

Farb, Peter, and the Editors of Time-Life Books. 1970. *Ecology*. New York: Time-Life Books.

Grissell, Eric. 2001. *Insects and Gardens: In Pursuit of a Garden Ecology*. Portland, Oregon: Timber Press.

Halpin, Anne. 1996. *Horticulture Gardener's Desk Reference*. New York: Macmillan.

Huxley, Anthony. 1998. *An Illustrated History of Gardening*. New York: The Lyons Press.

Leopold, Aldo. 1966. *A Sand County Almanac*. New York: Oxford University Press.

Moore-Landecker, Elizabeth. 1990. *Fundamentals of the Fungi*. 3rd ed. Englewood Cliffs, New Jersey: Prentice Hall.

Organic Gardening. Published six times per year. Emmaus, Pennsylvania: Rodale.

Proctor, Michael, Peter Yeo, and Andrew Lack. 1996. *The Natural History of Pollination*. Portland, Oregon: Timber Press.

Ricklefs, Robert E. 1997. *The Economy of Nature*. 4th ed. New York: W. H. Freeman and Co.

Further Reading

Smith, Miranda, and Anna Carr. 1988. *Rodale's Garden Insect, Disease, and Weed Identification Guide*. Emmaus, Pennsylvania: Rodale.

Staff of *Organic Gardening*. 1978. *The Encyclopedia of Organic Gardening*. Emmaus, Pennsylvania: Rodale.

Stern, Kingsley R., Shelley Jansky, and James E. Bidlack. 2003. *Introductory Plant Biology*. 9th ed. New York: McGraw-Hill.

Common and Scientific Name Index

Some broad groups of organisms, such as bacteria or birds, are listed in the Subject Index, and page references may be found there. Page references for line drawings are shown in **boldface**.

glass (*Ophisaurus*), 212
  green anole (*Anolis carolinensis*),
    plate 126
  horned toads (*Phrynosoma*), 56, 57
Lobelia, 56
locust, black (*Robinia pseudoacacia*),
    52, 145
locusts (Acridoidea), 53, 60, 65, 255
Lodoicea maldivica, 12
Lolium, 130
  annua, 174
Lonchocarpus, 20
Lonicera, 22
  japonica, 178; plate 15
loosestrife, purple (*Lythrum salicaria*),
    177, 183, 184, 273; plate 108
lotuses (*Nelumbo*), 123
  sacred (*Nelumbo nucifera*), 27, 177;
    plate 82
lousewort, common (*Pedicularis
    canadensis*), 42, 43
Loxodonta, 47, 51, 162
Loxosceles reclusa, 70
luffa, ridged (*Luffa acutangula*), plate
    76
Lumbricus rubellus, 73
  terrestris, 73
lupines (*Lupinus*), 13, 196
Lycaenidae, 212
Lycopersicon esculentum. See
    Lycopersicon lycopersicum
  lycopersicum, 37, 55, 64, 74, 124,
    125, 128, 133, 135, 151, 172, 183,
    188, 192, 197, 218, 221, 222, 250,
    258, 265; plates 89, 134, 164, 165
Lycosidae, 71
Lygodium microphyllum, 174; plate 104
Lygus, 65
  lineolaris, 62
Lymantria dispar, 62, 64, 186, 187, 197,
    253; plates 109, 142
Lysichiton, L. americanus, 123; plate 74

Lythrum salicaria, 177, 183, 184, 273;
    plate 108

Macrocystis, 167
maggot flies (*Rhagoletis*), 62
Magicicada, 66, 215, 251
magnolias (*Magnolia*), 12, 22; plate 94
maidenhair tree (*Ginkgo biloba*), 23
maize (*Zea mays*). See corn
Malacosoma, 64
  americanum, 236
mallards (*Anas platyrhynchos*), 54, 183
mallows (*Malva*), 51, 175, 222
Malus, 27, 36, 37, 64, 133, 145, 217, 221,
    222
Malva, 51, 175, 222
Manduca, 64, 255; plate 15
mango (*Mangifera indica*), 177
mangrove, red (*Rhizophora mangle*), 17
Manihot esculenta, 17
Manila hemp (*Musa textilis*), 14
Manilkara zapota, 16
manioc (*Manihot esculenta*), 17
manroots (*Marah*), 17
Mantidae, 69, 204, 207, 255
mantises, Central American
    (*Acanthops*), 211
  praying/preying (Mantidae), 69,
    204, 207, 255
Mantodea. See Mantidae
maples (*Acer*), 13, 28, 135, 178, 254
  silver (*Acer saccharinum*), 97
  sugar (*Acer saccharum*), 157
Marah, 17
Marasmius oreades, 98
marigolds (*Tagetes*), 38, 251, 259, 265
  French dwarf (*Tagetes patula*), 251
marjoram (*Origanum majorana*), 144
Marmota monax, 188, 202, 212
marsh marigold (*Caltha*), 138
martin, purple (*Progne subis*), 54, 246
Mastophora, 207

peaches (*Amygdalus persica*, *Prunus persica*), 37, 64, 127, 130, 133, 145, 153, 188, 220, 234; plates 79, 113

peanut (*Arachis hypogaea*), 27, 30

pears (*Pyrus*), 222

pecan (*Carya illinoinensis*), 138, 153

Pectobacterium carotovorum ssp. *atrosepticum*. See *Erwinia atroseptica*

Pedicularis canadensis, 42, 43
   groenlandica, plate 27

Pelargonium; plate 6

Penicillium, 82
   charlesii, plate 52

Penstemon, 21

peppers (*Capsicum*), 74, 124, 125, 128, 133, 254
   bell (*Capsicum annuum*), 258; plates 81, 136
   chili (*Capsicum annuum*), 254

pepperweed (*Lepidium virginicum*), 179

periwinkles (*Vinca*), 138

common (*Vinca minor*), 12

Peromyscus, plate 101

Peronospora, 221

Peronosporales, 83

Persea americana, 29, 221

Persian shield (*Strobilanthes dyerianus*), 13

Persicaria hydropiper, 250

Petroselinum sativum, 15

Petunia ×*hybrida*, 34

Phalangiidae, 70, 213; plate 46

Phascolarctos cinereus, 201, 202

Phaseolus, 27, **31**, 133, 136, 139, 145
   limensis, 143
   lunatus, 143
   vulgaris, 27, 30, 133, 136, 139

Phasmatidae, 211

Pheucticus melanocephalus, 213

Philaenus spumarius, 190

philodendrons (*Philodendron*), 17, 132

Phocoenidae, 205

Phoradendron, 225

phorid flies (Phoridae), 67

Phormia, 67

Photuris, 211, 254

Phrynosoma, 56, 57

Phyllobates terribilis, 213

Phyllonorycter crataegella, 62

Phyllostachys, 35

Phymata, 66

Physalis ixocarpa, 180

Phytolacca americana, 173, 178

phytophthora (*Phytophthora*), 221. See also fungi
   infestans, 221

Phytoseiidae, 71

Phytoseiulus, 257

Picea, 156

Piciformes, 55

Picoides pubescens, **53**

Pieris brassicae, 197
   rapae, 64, 108

pigeons (Columbiformes), 55

pigweeds (*Amaranthus albus*, *A. retroflexus*), 173, 176

pill bugs (Isopoda), 48, 58, 73

pine, Norfolk (*Araucaria heterophylla*), 96

pines (*Pinus*), 39, 187; plate 25
   eastern white (*Pinus strobus*), 157, 172, 217
   jack (*Pinus banksiana*), 160, 181; plate 96

pinks (*Dianthus*), 138

Piper methysticum, 20

Pisum sativum, 30, 31, 151, 191, 227

pitcher plants (*Nepenthes*, *Sarracenia*), 14

Pitohui, 214

Pituophis catenifer sayi, 56
   melanoleucus. See *Pituophis catenifer sayi*

Venus flytrap (*Dionaea muscipula*), 14,
140, 204; plate 7
Verbascum lynchnitis, 44
nigrum, 44
×*schiedeanum*, 44
thapsus, 177
Vernonia fasciculata, 176
Verticillium, 75, 220, 221, 262; plates
136, 164
Vespula, 68, 255
vetch, crown (*Coronilla varia*), 179
Viburnum, 246
lantanoides, plate 3
Vinca minor, 12
vincas (*Vinca*), 138
Viola ×*wittrockiana*, 128
violets (*Viola*), 39, 124
bird's foot (*Viola pedata*), 30
Virginia valeriae, plate 33
virus, plum necrosis, 220
plum pox, plate 135
potato spindle tuber, plate 140
Viscum, 225
Vitis, 12, 27, 36, 37, 192, 234; plates 91,
141
vultures, New World (Cathartidae), 54
Old World (Accipitridae), 54

walking sticks (Phasmatidae), 211
northern (*Diapheromera femorata*),
plate 129
walnut, black (*Juglans nigra*), 172
English (*Juglans regia*), 138
warbler, yellow-rumped (*Dendroica
coronata*), plate 155
wasps (Hymenoptera), 67, 68, 205, 206,
213, 214, 224, 247; plate 159
Anagrus epos, 248
braconid/brachonid (Braconidae,
Cotesia glomerata), 68, 197, 255
caprifig (Agaonidae, Torymidae), 235
chalcid (Chalcidoidae), 255

Encarsia formosa, 255; plate 162
ichneumonid (Ichneumonidae), 68,
255
jewel. See chalcid
short-tailed ichneumons (*Ophion*),
plate 142
trichogrammatid
(Trichogrammatidae), 68
vespid (*Paravespula, Vespula*), 68,
255
water bears. See tardigrades
watercress (*Nasturtium officinale*), 142
waterfowl (Anseriformes), 54
water hyacinth (*Eichhornia crassipes*),
plate 69
water lilies (*Nymphaea*), 142
water meal (*Wolffia*), 12
watermelon (*Citrullus lanatus*), 125
water pepper (*Persicaria hydropiper*), 250
waterweed (*Elodea*), 38
watery soft rot (*Sclerotinia sclerotiorum*), 221. See also fungi
waxwing, Bohemian (*Bombycilla garrulus*), 193
cedar (*Bombycilla cedrorum*), plate
155
weasels (*Mustela*), 209
webworms, sod (*Crambus*), 64
weevils (Curculionidae), 48, 64, 193;
plate 118
black vine (*Otiorhynchus sulcatus*), 255
maize (*Sitophilus zeamaise*), plate 118
whales, baleen (Mysticeti), 236
wheat (*Triticum*), 28
bread (*Triticum aestivum*), 42, 144
whip-poor-will (*Caprimulgus
vociferus*), 55
whiteflies (Aleyrodidae), 62, 66, 190,
224, 255
greenhouse (*Trialeurodes vaporariorum*), 66
wild olive family. See Elaeagnaceae

Subject Index

Page references for line drawings are shown in **boldface**.